EMILY FOX-SETON

Emily Fox-Seton

FRANCES HODGSON BURNETT

EMILY FOX-SETON

Being
"The Making of a Marchioness"
and
"The Methods of Lady Walderhurst"

ILLUSTRATED BY
C. D. WILLIAMS

BIBLIOBAZAAR

EMILY FOX-SETON

CONTENTS

PART ONE

PART TWO

PART ONE

CHAPTER ONE

When Miss Fox-Seton descended from the twopenny bus as it drew up, she gathered her trim tailor-made skirt about her with neatness and decorum, being well used to getting in and out of twopenny buses and to making her way across muddy London streets. A woman whose tailor-made suit must last two or three years soon learns how to protect it from splashes, and how to aid it to retain the freshness of its folds. During her trudging about this morning in the wet, Emily Fox-Seton had been very careful, and, in fact, was returning to Mortimer Street as unspotted as she had left it. She had been thinking a good deal about her dress— this particular faithful one which she had already worn through a twelvemonth. Skirts had made one of their appalling changes, and as she walked down Regent Street and Bond Street she had stopped at the windows of more than one shop bearing the sign "Ladies' Tailor and Habit-Maker," and had looked at the tautly attired, preternaturally slim models, her large, honest hazel eyes wearing an anxious expression. She was trying to discover *where* seams were to be placed and how gathers were to be hung; or if there were to be gathers at all; or if one had to be bereft of every seam in a style so unrelenting as to forbid the possibility of the honest and semi-penniless struggling with the problem of remodelling last season's skirt at all. "As it is only quite an ordinary brown," she had murmured to herself, "I might be able to buy a yard or so to match it, and I *might* be able to join the gore near the pleats at the back so that it would not be seen."

She quite beamed as she reached the happy conclusion. She was such a simple, normal-minded creature that it took but little to brighten the aspect of life for her and to cause her to break into her good-natured, childlike smile. A little kindness from any one, a little pleasure or a little comfort, made her glow with nice-tempered enjoyment. As she got out of the bus, and picked up her rough brown skirt, prepared to tramp bravely through the mud of Mortimer Street to her lodgings, she was positively radiant. It was not only her smile which was childlike, her face itself was childlike for a woman of her age and size. She was thirty-four and a well-set-up creature, with fine square shoulders and a long small waist and good hips. She was a big woman, but carried herself well, and having solved the problem of obtaining, through marvels of energy and management, one good dress a year, wore it so well, and changed her old ones so dexterously, that she always looked rather smartly dressed. She had nice, round, fresh cheeks and nice, big, honest eyes, plenty of mouse-brown hair and a short, straight nose. She was striking and well-bred-looking, and her plenitude of good-natured interest in everybody, and her pleasure in everything out of which pleasure could be wrested, gave her big eyes a fresh look which made her seem rather like a nice overgrown girl than a mature woman whose life was a continuous struggle with the narrowest of mean fortunes.

She was a woman of good blood and of good education, as the education of such women goes. She had few relatives, and none of them had any intention of burdening themselves with her pennilessness. They were people of excellent family, but had quite enough to do to keep their sons in the army or navy and find husbands for their daughters. When Emily's mother had died and her small annuity had died with her, none of them had wanted the care of a big raw-boned girl, and Emily had had the situation frankly explained to her. At eighteen she had begun to work as assistant teacher in a small school; the year following she had taken a place as nursery-governess; then she had been reading-companion to an unpleasant old woman in Northumberland. The old woman had lived in the country, and her relatives had hovered over her like vultures awaiting her decease. The household had been gloomy and gruesome enough to have driven into melancholy madness any girl not of the sanest and most matter-of-fact temperament. Emily Fox-Seton had endured it with an unfailing good nature, which at last had actually awakened in the breast of her mistress a ray of human feeling. When the old woman at length died, and Emily was to be turned out into the world, it was revealed that she had been left a legacy of a few hundred pounds, and a letter containing some rather practical, if harshly expressed, advice.

> Go back to London [Mrs. Maytham had written in her feeble, crabbed hand]. You are not clever enough to do anything remarkable in the way of earning your living, but you are so good-natured that you can make yourself useful to a lot of helpless creatures who will pay you a trifle for looking after them and the affairs they are too lazy or too foolish to manage for themselves. You might get on to one of the second-class fashion-papers to answer ridiculous questions about house-keeping or wall-papers or freckles. You know the kind of thing I mean. You might write notes or do accounts and shopping for some lazy woman. You are a practical, honest creature, and you have good manners. I have often thought that you had just the kind of commonplace gifts that a host of commonplace people want to find at their service. An old servant of mine who lives in Mortimer Street would probably give you cheap, decent lodgings, and behave well to you for my sake. She has reason to be fond of me. Tell her I sent you to her, and that she must take you in for ten shillings a week.

Emily wept for gratitude, and ever afterward enthroned old Mrs. Maytham on an altar as a princely and sainted benefactor, though after she had invested her legacy she got only twenty pounds a year from it.

"It was so *kind* of her," she used to say with heartfelt humbleness of spirit. "I never *dreamed* of her doing such a generous thing. I hadn't a *shadow* of a claim upon her—not a *shadow*." It was her way to express her honest emotions with emphasis which italicised, as it were, her outpourings of pleasure or appreciation.

She returned to London and presented herself to the ex-serving-woman. Mrs. Cupp had indeed reason to remember her mistress gratefully. At a time when youth and indiscreet affection had betrayed her disastrously, she had been saved from open disgrace and taken care of by Mrs. Maytham.

The old lady, who had then been a vigorous, sharp-tongued, middle-aged woman, had made the soldier lover marry his despairing sweetheart, and when he had promptly drunk himself to death, she had set her up in a lodging-house which had thriven and enabled her to support herself and her daughter decently.

In the second story of her respectable, dingy house there was a small room which she went to some trouble to furnish up for her dead mistress's friend. It was made into a bed-sitting-room with the aid of a cot which Emily herself bought and disguised decently as a couch during the daytime, by means of a red and blue Como blanket. The one window of the room looked out upon a black little back-yard and a sooty wall on which thin cats crept stealthily or sat and mournfully gazed at fate. The Como rug played a large part in the decoration of the apartment. One of them, with a piece of tape run through a hem, hung over the door in the character of a *portière*; another covered a corner which was Miss Fox-Seton's sole wardrobe. As she began to get work, the cheerful, aspiring creature bought herself a Kensington carpet-square, as red as Kensington art would permit it to be. She covered her chairs with Turkey-red cotton, frilling them round the seats. Over her cheap white muslin curtains (eight and eleven a pair at Robson's) she hung Turkey-red draperies. She bought a cheap cushion at one of Liberty's sales, and some bits of twopenny-halfpenny art china for her narrow mantelpiece. A lacquered tea-tray and a tea-set of a single cup and saucer, a plate and a teapot, made her feel herself almost sumptuous. After a day spent in trudging about in the wet or cold of the streets, doing other people's shopping, or searching for dressmakers or servants' characters for her patrons, she used to think of her bed-sitting-room with joyful anticipation. Mrs. Cupp always had a bright fire glowing in her tiny grate when she came in, and when her lamp was lighted under its home-made shade of crimson Japanese paper, its cheerful air, combining itself with the singing of her little, fat, black kettle on the hob, seemed absolute luxury to a tired, damp woman.

Mrs. Cupp and Jane Cupp were very kind and attentive to her. No one who lived in the same house with her could have helped liking her. She gave so little trouble, and was so expansively pleased by any attention, that the Cupps,—who were sometimes rather bullied and snubbed by the "professionals" who generally occupied their other rooms,—quite loved her. Sometimes the "professionals," extremely smart ladies and gentlemen who did turns at the balls or played small parts at theatres, were irregular in their payments or went away leaving bills behind them; but Miss Fox-Seton's payments were as regular as Saturday night, and, in fact, there had been times when, luck being against her, Emily had gone extremely hungry during a whole week rather than buy her lunches at the ladies' tea-shops with the money that would pay her rent.

In the honest minds of the Cupps, she had become a sort of possession of which they were proud. She seemed to bring into their dingy lodging-house a touch of the great world,—that world whose people lived in Mayfair and had country-houses where they entertained parties for the shooting and the hunting, and in which also existed the maids and matrons who on cold spring mornings sat, amid billows of satin and tulle and lace, surrounded with nodding plumes, waiting, shivering, for hours in their carriages that they might at last enter Buckingham Palace and be admitted to the Drawing-room. Mrs. Cupp knew that Miss Fox-Seton was "well connected;" she knew that she possessed an aunt with a title, though her ladyship never took the slightest notice of her niece. Jane Cupp took "Modern Society," and now and then had the pleasure of reading aloud to her young man little incidents concerning some castle or manor in which Miss Fox-Seton's aunt, Lady Malfry, was staying with earls and special favorites of the Prince's. Jane also knew that Miss Fox-Seton occasionally sent letters addressed "To the Right Honourable the Countess of So-and-so," and received replies stamped with coronets. Once even a letter had arrived adorned with strawberry-leaves, an incident which Mrs. Cupp and Jane had discussed with deep interest over their hot buttered-toast and tea.

Emily Fox-Seton, however, was far from making any professions of grandeur. As time went on she had become fond enough of the Cupps to be quite frank with them about her connections with these grand people. The countess had heard from a friend that Miss Fox-Seton had once found her an excellent governess, and she had commissioned her to find for her a reliable young ladies' serving-maid. She had done some secretarial work for a charity of which the duchess was patroness. In fact, these people knew her only as a well-bred woman who for a modest remuneration would make herself extremely useful in numberless practical ways. She knew much more of them than they knew of her, and, in her affectionate admiration for those who treated her with human kindness, sometimes spoke to Mrs. Cupp or Jane of their beauty or charity with a very nice, ingenuous feeling. Naturally some of her patrons grew fond of her, and as she was a fine, handsome young woman ackground with a perfectly correct bearing, they gave her little pleasures, inviting her to tea or luncheon, or taking her to the theatre.

Her enjoyment of these things was so frank and grateful that the Cupps counted them among their own joys. Jane Cupp—who knew something of dressmaking—felt it a brilliant thing to be called upon to renovate an old dress or help in the making of a new one for some festivity. The Cupps thought their tall, well-built lodger something of a beauty, and when they had helped her to dress for the evening, baring her fine, big white neck and arms, and adorning her thick braids of hair with some sparkling, trembling ornaments, after putting her in her four-wheeled cab, they used to go back to their kitchen and talk about her, and wonder that some gentleman who wanted a handsome, stylish woman at the head of his table, did not lay himself and his fortune at her feet.

"In the photograph-shops in Regent Street you see many a lady in a coronet that hasn't half the good looks she has," Mrs. Cupp remarked frequently. "She's got a nice complexion and a fine head of hair, and—if you ask *me*—she's got as nice a pair of clear eyes as a lady could have. Then look at her figure—her neck and her waist! That kind of big long throat of hers would set off rows of pearls or diamonds beautiful! She's a lady born, too, for all her simple, every-day way; and she's a sweet creature, if ever there was one. For kind-heartedness and good-nature I never saw her equal."

Miss Fox-Seton had middle-class patrons as well as noble ones,—in fact, those of the middle class were far more numerous than the duchesses,—so it had been possible for her to do more than one good turn for the Cupp household. She had got sewing in Maida Vale and Bloomsbury for Jane Cupp many a time, and Mrs. Cupp's dining-room floor had been occupied for years by a young man Emily had been able to recommend. Her own appreciation of good turns made her eager to do them for others. She never let slip a chance to help any one in any way.

It was a good-natured thing done by one of her patrons who liked her, which made her so radiant as she walked through the mud this morning. She was inordinately fond of the country, and having had what she called "a bad winter," she had not seen the remotest chance of getting out of town at all during the summer months. The weather was beginning to be unusually hot, and her small red room, which seemed so cosy in winter, was shut in by a high wall from all chance of breezes. Occasionally she lay and panted a little in her cot, and felt that when all the private omnibuses, loaded with trunks and servants, had rattled away and deposited their burdens at the various stations, life in town would be rather lonely. Every one she knew would have gone somewhere, and Mortimer Street in August was a melancholy thing.

And Lady Maria had actually invited her to Mallowe. What a piece of good fortune-what an extraordinary piece of kindness!

She did not know what a source of entertainment she was to Lady Maria, and how the shrewd, worldly old thing liked her. Lady Maria Bayne was the cleverest, sharpest-tongued, smartest old woman in London. She knew everybody and had done everything in her youth, a good many things not considered highly proper. A certain royal duke had been much pleased with her and people had said some very nasty things about it. But this had not hurt Lady Maria. She knew how to say nasty things herself, and as she said them wittily they were usually listened to and repeated.

Emily Fox-Seton had gone to her first to write notes for an hour every evening. She had sent, declined, and accepted invitations, and put off charities and dull people. She wrote a fine, dashing hand, and had a matter-of-fact intelligence and knowledge of things. Lady Maria began to depend on her and to find that she could be sent on errands and depended on to do a number of things. Consequently, she was often at South Audley Street, and once, when Lady Maria was suddenly taken ill and was horribly frightened about herself, Emily was such a comfort to her that she kept her for three weeks.

"The creature is so cheerful and perfectly free from vice that she's a relief," her ladyship said to her nephew afterward. "So many women are affected cats. She'll go out and buy you a box of pills or a porous plaster, but at the same time she has a kind of simplicity and freedom from spites and envies which might be the natural thing for a princess."

So it happened that occasionally Emily put on her best dress and most carefully built hat and went to South Audley Street to tea. (Sometimes she had previously gone in buses to some remote place in the City to buy a special tea of which there had been rumours.) She met some very smart people and rarely any stupid ones, Lady Maria being incased in a perfect, frank armour of good-humoured selfishness, which would have been capable of burning dulness at the stake.

"I won't have dull people," she used to say. "I'm dull myself."

When Emily Fox-Seton went to her on the morning in which this story opens, she found her consulting her visiting-book and making lists.

"I'm arranging my parties for Mallowe," she said rather crossly. "How tiresome it is! The people one wants at the same time are always nailed to the opposite ends of the earth. And then things are found out about people, and one can't have them till it's blown over. Those ridiculous Dexters! They were the nicest possible pair—both of them good-looking and both of them ready to flirt with anybody. But there was too much flirting, I suppose. Good heavens! if I couldn't have a scandal and keep it quiet, I wouldn't have a scandal at all. Come and help me, Emily."

Emily sat down beside her.

"You see, it is my early August party," said her ladyship, rubbing her delicate little old nose with her pencil, "and Walderhurst is coming to me. It always amuses me to have Walderhurst. The moment a man like that comes into a room the women begin to frisk about and swim and languish, except those who try to get up interesting conversations they think likely to attract his attention. They all think it is possible that he may marry them. If he were a Mormon he might have marchionesses of Walderhurst of all shapes and sizes."

"I suppose," said Emily, "that he was very much in love with his first wife and will never marry again."

"He wasn't in love with her any more than he was in love with his housemaid. He knew he must marry, and thought it very annoying. As the child died, I believe he thinks it his duty to marry again. But he hates it. He's rather dull, and he can't bear women fussing about and wanting to be made love to."

They went over the visiting-book and discussed people and dates seriously. The list was made and the notes written before Emily left the house. It was not until she had got up and was buttoning her coat that Lady Maria bestowed her boon.

"Emily," she said, "I am going to ask you to Mallowe on the 2d. I want you to help me to take care of people and keep them from boring me and one another, though I don't mind their boring one another half so much as I mind their boring me. I want to be able to go off and take my nap at any hour I choose. I will *not* entertain people. What you can do is to lead them off to gather things of look at church towers. I hope you'll come."

Emily Fox-Seton's face flushed rosily, and her eyes opened and sparkled.

"O Lady Maria, you *are* kind!" she said. "You know how I should enjoy it. I have heard so much of Mallowe. Every one says it is so beautiful and that there are no such gardens in England."

"They are good gardens. My husband was rather mad about roses. The best train for you to take is the 2:30 from Paddington. That will bring you to the Court just in time for tea on the lawn."

Emily could have kissed Lady Maria if they had been on the terms which lead people to make demonstrations of affection. But she would have been quite as likely to kiss the butler when he bent over her at dinner and murmured in dignified confidence, "Port or sherry, miss?" Bibsworth would have been no more astonished than Lady Maria would, and Bibsworth certainly would have expired of disgust and horror.

She was so happy when she hailed the twopenny bus that when she got into it her face was beaming with the delight which adds freshness and good looks to any woman. To think that such good luck had come to her! To think of leaving her hot little room behind her and going as a guest to one of the most beautiful old houses in England! How delightful it would be to live for a while quite naturally the life the fortunate people lived year after year—to be a part of the beautiful order and picturesqueness and dignity of it! To sleep in a lovely bedroom, to be called in the morning by a perfect housemaid, to have one's early tea served in a delicate cup, and to listen as one drank it to the birds singing in the trees in the park! She had an ingenuous appreciation of the simplest material joys, and the fact that she would wear her nicest clothes every day, and dress for dinner every evening, was a delightful thing to reflect upon. She got so much more out of life than most people, though she was not aware of it.

She opened the front door of the house in Mortimer Street with her latch-key, and went upstairs, almost unconscious that the damp heat was dreadful. She met Jane Cupp coming down, and smiled at her happily.

"Jane," she said, "if you are not busy, I should like to have a little talk with you. Will you come into my room?"

"Yes, miss," Jane replied, with her usual respectful lady's maid's air. It was in truth Jane's highest ambition to become some day maid to a great lady, and she privately felt that her association with Miss Fox-Seton was the best possible training. She used to ask to be allowed to dress her when she went out, and had felt it a privilege to be permitted to "do" her hair.

She helped Emily to remove her walking dress, and neatly folded away her gloves and veil. She knelt down before her as soon as she saw her seat herself to take off her muddy boots.

"Oh, *thank* you, Jane," Emily exclaimed, with her kind italicised manner. "That *is* good of you. I *am* tired, really. But such a nice thing has happened. I have had such a delightful invi-tation for the first week in August."

"I'm sure you'll enjoy it, miss," said Jane. "It's so hot in August."

"Lady Maria Bayne has been kind enough to invite me to Mallowe Court," explained Emily, smiling down at the cheap slipper Jane was putting on her large, well-shaped foot. She was built on a large scale, and her foot was of no Cinderella-like proportions.

"O miss!" exclaimed Jane. "How beautiful! I was reading about Mallowe in 'Modern Society' the other day, and it said it was lovely and her ladyship's parties were wonderful for smartness. The paragraph was about the Marquis of Walderhurst."

"He is Lady Maria's cousin," said Emily, "and he will be there when I am."

She was a friendly creature, and lived a life so really isolated from any ordinary companionship that her simple little talks with Jane and Mrs. Cupp were a pleasure to her. The Cupps were neither gossiping nor intrusive, and she felt as if they were her friends. Once when she had been ill for a week she remembered suddenly realising that she had no intimates at all, and that if she died Mrs. Cupp's and Jane's would certainly be the last faces—and the only ones—she would see. She had cried a little the night she thought of it, but then, as she told herself, she was feverish and weak, and it made her morbid.

"It was because of this invitation that I wanted to talk to you, Jane," she went on. "You see, we shall have to begin to contrive about dresses."

"Yes, indeed, miss. It's fortunate that the summer sales are on, isn't it? I saw some beautiful colored linens yesterday. They were so cheap, and they do make up so smart for the country. Then you've got your new Tussore with the blue collar and waistband. It does become you."

"I must say I think that a Tussore always looks fresh," said Emily, "and I saw a really nice little tan toque—one of those soft straw ones—for three and eleven. And just a twist of blue chiffon and a wing would make it look quite *good*."

She was very clever with her fingers, and often did excellent things with a bit of chiffon and a wing, or a few yards of linen or muslin and a remnant of lace picked up at a sale. She and Jane spent quite a happy afternoon in careful united contemplation of the resources of her limited wardrobe. They found that the brown skirt *could* be altered, and, with the addition of new *revers* and collar and a *jabot* of string-coloured lace at the neck, would look quite fresh. A black net evening dress, which a patron had good-naturedly given her the year before, could be remodelled and touched up delightfully. Her fresh face and her square white shoulders were particularly adorned by black. There was a white dress which could be sent to the cleaner's, and an old pink one whose superfluous breadths could be combined with lace and achieve wonders.

"Indeed, I think I shall be very well off for dinner-dresses," said Emily. "Nobody expects me to change often. Every one knows—if they notice at all." She did not know she was humble-minded and of an angelic contentedness of spirit. In fact, she did not find herself interested in contemplation of her own qualities, but in contemplation and admiration of those of other people. It was necessary to provide Emily Fox-Seton with food and lodging and such a wardrobe as would be just sufficient credit to her more fortunate acquaintances. She worked hard to attain this modest end and was quite satisfied. She found at the shops where the summer sales were being held a couple of cotton frocks to which her height and her small, long waist gave an air of actual elegance. A sailor hat, with a smart ribbon and well-set quill, a few new trifles for her neck, a bow, a silk handkerchief daringly knotted, and some fresh gloves, made her feel that she was sufficiently equipped.

During her last expedition to the sales she came upon a nice white duck coat and skirt which she contrived to buy as a present for Jane. It was necessary to count over the contents of her purse very carefully and to give up the purchase of a slim umbrella she wanted, but she did it cheerfully. If she had been a rich woman she would have given presents to every one she knew, and it was actually a luxury to her to be able to do something for the Cupps, who, she always felt, were continually giving her more than she paid for. The care they took of her small room, the fresh hot tea they managed to have ready when she came in, the penny bunch of daffodils they sometimes put on her table, were kindnesses, and she was grateful for them. "I am very much obliged to you, Jane," she said to the girl, when she got into the four-wheeled cab on the eventful day of her journey to Mallowe. "I don't know what I should have done without you, I'm sure. I feel so smart in my dress now that you have altered it. If Lady Maria's maid ever thinks of leaving her, I am sure I could recommend you for her place."

CHAPTER TWO

There were other visitors to Mallowe Court travelling by the 2:30 from Paddington, but they were much smarter people than Miss Fox-Seton, and they were put into a first-class carriage by a footman with a cockade and a long drab coat. Emily, who traveled third with some workmen with bundles, looked out of her window as they passed, and might possibly have breathed a faint sigh if she had not felt in such buoyant spirits. She had put on her revived brown skirt and a white linen blouse with a brown dot on it. A soft brown silk tie was knotted smartly under her fresh collar, and she wore her new sailor hat. Her gloves were brown, and so was her parasol. She looked nice and taut and fresh, but notably inexpensive. The people who went to sales and bought things at three and eleven or "four-three" a yard would have been able add her up and work out her total. But there would be no people capable of the calculation at Mallowe. Even the servants' hall was likely to know less of prices than this one guest did. The people the drab-coated footman escorted to the first-class carriage were a mother and daughter. The mother had regular little features, and would have been pretty if she had not been much too plump. She wore an extremely smart travelling-dress and a wonderful dust-cloak of cool, pale, thin silk. She was not an elegant person, but her appointments were luxurious and self-indulgent. Her daughter was pretty, and had a slim, swaying waist, soft pink cheeks, and a pouting mouth. Her large picture-hat of pale-blue straw, with its big gauze bow and crushed roses, had a slightly exaggerated Parisian air.

"It is a little too picturesque," Emily thought; "but how lovely she looks in it! I suppose it was so becoming she could not help buying it. I'm sure it's Virot."

As she was looking at the girl admiringly, a man passed her window. He was a tall man with a square face. As he passed close to

Emily, he stared through her head as if she had been transparent or invisible. He got into the smoking-carriage next to her.

When the train arrived at Mallowe station, he was one of the first persons who got out. Two of Lady Maria's men were waiting on the platform. Emily recognised their liveries. One met the tall man, touching his hat, and followed him to a high cart, in the shafts of which a splendid iron-gray mare was fretting and dancing. In a few moments the arrival was on the high seat, the footman behind, and the mare speeding up the road. Miss Fox-Seton found herself following the second footman and the mother and daughter, who were being taken to the landau waiting outside the station. The footman piloted them, merely touching his hat quickly to Emily, being fully aware that she could take care of herself.

This she did promptly, looking after her box, and seeing it safe in the Mallowe omnibus. When she reached the landau, the two other visitors were in it. She got in, and in entire contentment sat down with her back to the horses.

The mother and daughter wore for a few minutes a somewhat uneasy air. They were evidently sociable persons, but were not quite sure how to begin a conversation with an as yet unintroduced lady who was going to stay at the country house to which they were themselves invited.

Emily herself solved the problem, producing her commonplace with a friendly tentative smile.

"Isn't it a lovely country?" she said.

"It's perfect," answered the mother. "I've never visited Europe before, and the English country seems to me just exquisite. We have a summer place in America, but the country is quite different."

She was good-natured and disposed to talk, and, with Emily Fox-Seton's genial assistance, conversation flowed. Before they were half-way to Mallowe, it had revealed itself that they were from Cincinnati, and after a winter spent in Paris, largely devoted to visits to Paquin, Doucet, and Virot, they had taken a house in Mayfair for the season. Their name was Brooke. Emily thought she remembered hearing of them as people who spent a great deal of money and went incessantly to parties, always in new and lovely clothes. The girl had been presented by the American minister, and had had a sort of success because she dressed and danced exquisitely. She was the kind of American girl who ended by marrying a title. She had sparkling eyes and a delicate tip-tilted nose. But even Emily guessed that she was an astute little person.

"Have you ever been to Mallowe Court before?" she inquired.

"No; and I am *so* looking forward to it. It is so beautiful."

"Do you know Lady Maria very well?"

"I've known her about three years. She has been very kind to me."

"Well, I shouldn't have taken her for a particularly kind person. She's too sharp."

Emily amiably smiled. "She's so clever," she replied.

"Do you know the Marquis of Walderhurst?" asked Mrs. Brooke.

"No," answered Miss Fox-Seton. She had no part in that portion of Lady Maria's life which was illumined by cousins who were marquises. Lord Walderhurst did not drop in to afternoon tea. He kept himself for special dinner-parties.

"Did you see the man who drove away in the high cart?" Mrs. Brooke continued, with a touch of fevered interest. "Cora thought it must be the marquis. The servant who met him wore the same livery as the man up there"—with a nod toward the box.

"It was one of Lady Maria's servants," said Emily; "I have seen him in South Audley Street. And Lord Walderhurst was to be at Mallowe. Lady Maria mentioned it."

"There, mother!" exclaimed Cora.

"Well, of course if he is to be there, it will make it interesting," returned her mother, in a tone in which lurked an admission of relief. Emily wondered if she had wanted to go somewhere else and had been firmly directed toward Mallowe by her daughter.

"We heard a great deal of him in London this season," Mrs. Brooks went on.

Miss Cora Brooke laughed.

"We heard that at least half a dozen people were determined to marry him," she remarked with pretty scorn. "I should think that to meet a girl who was indifferent might be good for him."

"Don't be too indifferent, Cora," said her mother, with ingenuous ineptness.

It was a very stupid bit of revelation, and Miss Brooke's eyes flashed. If Emily Fox-Seton had been a sharp woman, she would have observed that, if the *rôle* of indifferent and piquant young person could be made dangerous to Lord Walderhurst, it would be made so during this visit. The man was in peril from this beauty from Cincinnati and her rather indiscreet mother, though upon the whole, the indiscreet maternal parent might unconsciously form his protection.

But Emily only laughed amiably, as at a humorous remark. She was ready to accept almost anything as humour.

"Well, he *would* be a great match for any girl," she said. "He is so rich, you know. He is very rich."

When they reached Mallowe, and were led out upon the lawn, where the tea was being served under embowering trees, they found a group of guests eating little hot cakes and holding teacups in their hands. There were several young women, and one of them—a very tall, very fair girl, with large eyes as blue as forget-me-nots, and with a lovely, limp, and long blue frock of the same shade—had been one of the beauties of the past season. She was a Lady Agatha Slade, and Emily began to admire her at once. She felt her to be a sort of added boon bestowed by kind Fate upon herself. It was so delightful that she should be of this particular house-party—this lovely creature, whom she had only known previously through pictures in ladies' illustrated papers. If it should occur to her to wish to become the Marchioness of Walderhurst, what could possibly prevent the consummation of her desire? Surely not Lord Walderhurst himself, if he was human. She was standing, leaning lightly against the trunk of an ilex-tree, and a snow-white Borzoi was standing close to her, resting his long, delicate head against her gown, encouraging the caresses of her fair, stroking hand. She was in this attractive pose when Lady Maria turned in her seat and said:

"There's Walderhurst."

The man who had driven himself over from the station in the cart was coming towards them across the grass. He was past middle life and plain, but was of good height and had an air. It was perhaps, on the whole, rather an air of knowing what he wanted.

Emily Fox-Seton, who by that time was comfortably seated in a cushioned basket-chair, sipping her own cup of tea, gave him the benefit of the doubt when she wondered if he was not really distinguished and aristocratic-looking. He was really neither, but was well-built and well-dressed, and had good grayish-brown eyes, about the colour of his grayish-brown hair. Among these amiably worldly people, who were not in the least moved by an altruistic prompting, Emily's greatest capital consisted in the fact that she did not expect to be taken the least notice of. She was not aware that it was her capital, because the fact was so wholly a part of the simple contentedness of her nature that she had not thought about it at all. The truth was that she found all her entertainment and occupation in being an audience or a spectator. It did not occur to her to notice that, when the guests were presented to him, Lord Walderhurst barely glanced at her surface as he bowed, and could scarcely be said to forget her existence the next second, because he had hardly gone to the length of recognising it. As she enjoyed her extremely nice cup of tea and little buttered scone, she also enjoyed looking at his Lordship discreetly, and trying to make an innocent summing up of his mental attitudes.

Lady Maria seemed to like him and to be pleased to see him. He himself seemed, in an undemonstrative way, to like Lady Maria. He also was evidently glad to get his tea, and enjoyed it as he sat at his cousin's side. He did not pay very much attention to any one else. Emily was slightly disappointed to see that he did not glance at the beauty and the Borzoi more than twice, and then that his examination seemed as much for the Borzoi as for the beauty. She could not help also observing that since he had joined the circle it had become more animated, so far at least as the female members were concerned. She could not help remembering Lady Maria's remark about the effect he produced on women when he entered a room. Several interesting or sparkling speeches had already been made. There was a little more laughter and chattiness, which somehow it seemed to be quite open to Lord Walderhurst to enjoy, though it was not exactly addressed to him. Miss Cora Brooke, however, devoted herself to a young man in white flannels with an air of tennis about him. She sat a little apart and talked to him in a voice soft enough to even exclude Lord Walderhurst. Presently she and her companion got up and sauntered away. They went down the broad flight of ancient stone steps which led to the tennis-court, lying in full view below the lawn. There they began to play tennis. Miss Brooke skimmed and darted about like a swallow. The swirl of her lace petticoats was most attractive.

"That girl ought not to play tennis in shoes with ridiculous heels," remarked Lord Walderhurst. "She will spoil the court."

Lady Maria broke into a little chuckle.

"She wanted to play at this particular moment," she said. "And as she has only just arrived, it did not occur to her to come out to tea in tennis-shoes."

"She'll spoil the court all the same," said the marquis. "What clothes! It's amazing how girls dress now."

"I wish I had such clothes," answered Lady Maria, and she chuckled again. "She's got beautiful feet."

"She's got Louis Quinze heels," returned his Lordship.

At all events, Emily Fox-Seton thought Miss Brooke seemed to intend to rather keep out of his way and to practise no delicate allurements. When her tennis-playing was at an end, she sauntered about the lawn and terraces with her companion, tilting her parasol prettily over her shoulder, so that it formed an entrancing background to her face and head. She seemed to be entertaining the young man. His big laugh and the silver music of her own lighter merriment rang out a little tantalisingly.

"I wonder what Cora is saying," said Mrs. Brooke to the group at large. "She always makes men laugh so."

Emily Fox-Seton felt an interest herself, the merriment sounded so attractive. She wondered if perhaps to a man who had been so much run after a girl who took no notice of his presence and amused other men so much might not assume an agreeable aspect.

But he took more notice of Lady Agatha Slade than of any one else that evening. She was placed next to him at dinner, and she really was radiant to look upon in palest green chiffon. She had an exquisite little head, with soft hair piled with wondrous lightness upon it, and her long little neck swayed like the stem of a flower. She was lovely enough to arouse in the beholder's mind the anticipation of her being silly, but she was not silly at all.

Lady Maria commented upon that fact to Miss Fox-Seton when they met in her bedroom late that night. Lady Maria liked to talk and be talked to for half an hour after the day was over, and Emily Fox-Seton's admiring interest in all she said she found at once stimulating and soothing. Her Ladyship was an old woman who indulged and inspired herself with an Epicurean wisdom.

Though she would not have stupid people about her, she did not always want very clever ones.

"They give me too much exercise," she said. "The epigrammatic ones keep me always jumping over fences. Besides, I like to make all the epigrams myself."

Emily Fox-Seton struck a happy mean, and she was a genuine admirer. She was intelligent enough not to spoil the point of an epigram when she repeated it, and she might be relied upon to repeat it and give all the glory to its originator. Lady Maria knew there were people who, hearing your good things, appropriated them without a scruple. To-night she said a number of good things to Emily in summing up her guests and their characteristics.

"Walderhurst has been to me three times when I made sure that he would not escape without a new marchioness attached to him. I should think he would take one to put an end to the annoyance of dangling unplucked upon the bough. A man in his position, if he has character enough to choose, can prevent even his wife's being a nuisance. He can give her a good house, hang the family diamonds on her, supply a decent elderly woman as a sort of lady-in-waiting and turn her into the paddock to kick up her heels within the limits of decorum. His own rooms can be sacred to him. He has his clubs and his personal interests. Husbands and wives annoy each other very little in these days. Married life has become comparatively decent."

"I should think his wife might be very happy," commented Emily. "He looks very kind."

"I don't know whether he is kind or not. It has never been necessary for me to borrow money from him."

Lady Maria was capable of saying odd things in her refined little drawling voice.

"He's more respectable than most men of his age. The diamonds are magnificent, and he not only has three superb places, but has money enough to keep them up. Now, there are three aspirants at Mallowe in the present party. Of course you can guess who they are, Emily?"

Emily Fox-Seton almost blushed. She felt a little indelicate.

"Lady Agatha would be very suitable," she said. "And Mrs. Ralph is very clever, of course. And Miss Brooke is really pretty."

Lady Maria gave vent to her small chuckle.

"Mrs. Ralph is the kind of woman who means business. She'll corner Walderhurst and talk literature and roll her eyes at him until he hates her. These writing women, who are intensely pleased with themselves, if they have some good looks into the bargain, believe themselves capable of marrying any one. Mrs. Ralph has fine eyes and rolls them. Walderhurst won't be ogled. The Brooke girl is sharper than Ralph. She was very sharp this afternoon. She began at once."

"I—I didn't see her"—wondering.

"Yes, you did; but you didn't understand. The tennis, and the laughing with young Heriot on the terrace! She is going to be the piquant young woman who aggravates by indifference, and disdains rank and splendour; the kind of girl who has her innings in novelettes—but not out of them. The successful women are those who know how to toady in the right way and not obviously. Walderhurst has far too good an opinion of himself to be attracted by a girl who is making up to another man: he's not five-and-twenty."

Emily Fox-Seton was reminded, in spite of herself, of Mrs. Brooke's plaint: "Don't be too indifferent, Cora." She did not want to recall it exactly, because she thought the Brookes agreeable and would have preferred to think them disinterested. But, after all, she reflected, how natural that a girl who was so pretty should feel that the Marquis of Walderhurst represented prospects. Chiefly, however, she was filled with admiration at Lady Maria's cleverness.

"How wonderfully you observe everything, Lady Maria!" she exclaimed. "How wonderfully!"

"I have had forty-seven seasons in London. That's a good many, you know. Forty-seven seasons of débutantes and mothers tend toward enlightenment. Now there is Agatha Slade, poor girl! She's of a kind I know by heart. With birth and beauty, she is perfectly helpless. Her people are poor enough to be entitled to aid from the Charity Organisation, and they have had the indecency to present themselves with six daughters—six! All with delicate skins and delicate little noses and heavenly eyes. Most men can't afford them, and they can't afford most men. As soon as Agatha begins to go off a little, she will have to step aside, if she has not married. The others must be allowed their chance. Agatha has had the advertising of the illustrated papers this season, and she has gone well. In these days a new beauty is advertised like a new soap. They haven't given them sandwich-men in the streets, but that is about all that has been denied them. But Agatha has not had any special offer, and I know both she and her mother are a little frightened. Alix must come out next season, and they can't afford frocks for two. Agatha will have to be sent to their place in Ireland, and to be sent to Castle Clare is almost like being sent to the Bastille. She'll never get out alive. She'll have to stay there and see herself grow thin instead of slim, and colourless instead of fair. Her little nose will grow sharp, and she will lose her hair by degrees."

"Oh!" Emily Fox-Seton gave forth sympathetically. "What a pity that would be! I thought—I really thought—Lord Walderhurst seemed to admire her."

"Oh, every one admires her, for that matter; but if they go no further that will not save her from the Bastille, poor thing. There, Emily; we must go to bed. We have talked enough."

CHAPTER THREE

To awaken in a still, delicious room, with the summer morning sunshine breaking softly into it through leafy greenness, was a delightful thing to Miss Fox-Seton, who was accustomed to opening her eyes upon four walls covered with cheap paper, to the sound of outside hammerings, and the rattle and heavy roll of wheels. In a building at the back of her bed-sitting-room there lived a man whose occupation, beginning early in the morning, involved banging of a persistent nature.

She awakened to her first day at Mallowe, stretching herself luxuriously, with the smile of a child. She was so thankful for the softness of her lavender-fragrant bed, and so delighted with the lovely freshness of her chintz-hung room. As she lay upon her pillow, she could see the boughs of the trees, and hear the chatter of darting starlings. When her morning tea was brought, it seemed like nectar to her. She was a perfectly healthy woman, with a palate as unspoiled as that of a six-year-old child in the nursery. Her enjoyment of all things was so normal as to be in her day and time an absolute abnormality.

She rose and dressed at once, eager for the open air and sunshine. She was out upon the lawn before any one else but the Borzoi, which rose from beneath a tree and came with stately walk toward her. The air was exquisite, the broad, beautiful stretch of view lay warm in the sun, the masses of flowers on the herbaceous borders showed leaves and flower-cups adorned with glittering drops of dew. She walked across the spacious sweep of short-cropped sod, and gazed enraptured at the country spread out below. She could have kissed the soft white sheep dotting the fields and lying in gentle, huddled groups under the trees.

"The darlings!" she said, in a little, effusive outburst.

She talked to the dog and fondled him. He seemed to understand her mood, and pressed close against her gown when she stopped. They walked together about the gardens, and presently picked up an exuberant retriever, which bounded and wriggled and at once settled into a steady trot beside them. Emily adored the flowers as she walked by their beds, and at intervals stopped to bury her face in bunches of spicy things. She was so happy that the joy in her hazel eyes was pathetic.

She was startled, as she turned into a rather narrow rose-walk, to see Lord Walderhurst coming toward her. He looked exceedingly clean in his fresh light knickerbocker suit, which was rather becoming to him. A gardener was walking behind, evidently gathering roses for him, which he put into a shallow basket. Emily Fox-Seton cast about for a suitable remark to make, if he should chance to stop to speak to her. She consoled herself with the thought that there were things she really *wanted* to say about the beauty of the gardens, and certain clumps of heavenly-blue campanulas, which seemed made a feature of in the herbaceous borders. It was so much nicer not to be obliged to invent observations. But his lordship did not stop to speak to her. He was interested in his roses (which, she heard afterward, were to be sent to town to an invalid friend), and as she drew near, he turned aside to speak to the gardener. As Emily was just passing him when he turned again, and as the passage was narrow, he found himself unexpectedly gazing into her face.

Being nearly the same height, they were so near each other that it was a little awkward.

"I beg pardon," he said, stepping back a pace and lifting his straw hat.

But he did not say, "I beg pardon, Miss Fox-Seton," and Emily knew that he had not recognised her again, and had not the remotest idea who she was or where she came from.

She passed him with her agreeable, friendly smile, and there returned to her mind Lady Maria's remarks of the night before.

"To think that if he married poor pretty Lady Agatha she will be mistress of three places quite as beautiful as Mallowe, three lovely old houses, three sets of gardens, with thousands of flowers to bloom every year! How nice it would be for her! She is so lovely that it seems as if he *must* fall in love with her. Then, if

33

she was Marchioness of Walderhurst, she could do so much for her sisters."

After breakfast she spent her morning in doing a hundred things for Lady Maria. She wrote notes for her, and helped her to arrange plans for the entertainment of her visitors. She was very busy and happy. In the afternoon she drove across the moor to Maundell, a village on the other side of it. She really went on an errand for her hostess, but as she was fond of driving and the brown cob was a beauty, she felt that she was being given a treat on a level with the rest of her ladyship's generous hospitalities. She drove well, and her straight, strong figure showed to much advantage on the high seat of the cart. Lord Walderhurst himself commented on her as he saw her drive away.

"She has a nice, flat, straight back, that woman," he remarked to Lady Maria. "What is her name? One never hears people's names when one is introduced."

"Her name is Emily Fox-Seton," her ladyship answered, "and she's a nice creature."

"That would be an inhuman thing to say to most men, but if one is a thoroughly selfish being, and has some knowledge of one's own character, one sees that a nice creature might be a nice companion."

"You are quite right," was Lady Maria's reply, as she held up her lorgnette and watched the cart spin down the avenue. "I am selfish myself, and I realise that is the reason why Emily Fox-Seton is becoming the lodestar of my existence. There is such comfort in being pandered to by a person who is not even aware that she is pandering. She doesn't suspect that she is entitled to thanks for it."

That evening Mrs. Ralph came shining to dinner in amber satin, which seemed to possess some quality of stimulating her to brilliance. She was witty enough to collect an audience, and Lord Walderhurst was drawn within it. This was Mrs. Ralph's evening. When the men returned to the drawing-room, she secured his lordship at once and managed to keep him. She was a woman who could talk pretty well, and perhaps Lord Walderhurst was amused. Emily Fox-Seton was not quite sure that he was, but at least he listened. Lady Agatha Slade looked a little listless and pale. Lovely as she was, she did not always collect an audience, and this evening she said she had a headache. She actually crossed the room, and taking a seat by Miss Emily Fox-Seton, began to talk to her about Lady Maria's charity-knitting which she had taken up. Emily was so gratified that she found conversation easy. She did not realise that at that particular moment she was a most agreeable and comforting companion for Agatha Slade. She had heard so much of her beauty during the season, and remembered so many little things that a girl who was a thought depressed might like to hear referred to again. Sometimes to Agatha the balls where people had collected in groups to watch her dancing, the flattering speeches she had heard, the dazzling hopes which had been raised, seemed a little unreal, as if, after all, they could have been only dreams. This was particularly so, of course, when life had dulled for a while and the atmosphere of unpaid bills became heavy at home. It was so to-day, because the girl had received a long, anxious letter from her mother, in which much was said of the importance of an early preparation for the presentation of Alix, who had really been kept back a year, and was in fact nearer twenty than nineteen.

"If we were not in Debrett and Burke, one might be reserved about such matters," poor Lady Claraway wrote; "but what is one to do when all the world can buy one's daughters' ages at the book-sellers'?"

Miss Fox-Seton had seen Lady Agatha's portrait at the Academy and the way in which people had crowded about it. She had chanced to hear comments also, and she agreed with a number of persons who had not thought the picture did the original justice.

"Sir Bruce Norman was standing by me with an elderly lady the first time I saw it," she said, as she turned a new row of the big white-wool scarf her hostess was knitting for a Deep-Sea Fisherman's Charity. "He really looked quite annoyed. I heard him say: 'It is not good at all. She is far, far lovelier. Her eyes are like blue flowers.' The moment I saw you, I found myself looking at your eyes. I hope I didn't seem rude."

Lady Agatha smiled. She had flushed delicately, and took up in her slim hand a skein of the white wool.

"There are some people who are never rude," she sweetly said, "and you are one of them, I am sure. That knitting looks nice. I wonder if I could make a comforter for a deep-sea fisherman."

"If it would amuse you to try," Emily answered, "I will begin one for you. Lady Maria has several pairs of wooden needles. Shall I?"

"Do, please. How kind of you!"

In a pause of her conversation, Mrs. Ralph, a little later, looked across the room at Emily Fox-Seton bending over Lady Agatha and the knitting, as she gave her instructions.

"What a good-natured creature that is!" she said.

Lord Walderhurst lifted his monocle and inserted it in his unillumined eye. He also looked across the room. Emily wore the black evening dress which gave such opportunities to her square white shoulders and firm column of throat; the country air and sun had deepened the colour on her cheek, and the light of the nearest lamp fell kindly on the big twist of her nut-brown hair, and burnished it. She looked soft and warm, and so generously interested in her pupil's progress that she was rather sweet.

Lord Walderhurst simply looked at her. He was a man of but few words. Women who were sprightly found him somewhat unresponsive. In fact, he was aware that a man in his position need not exert himself. The women themselves would talk. They wanted to talk because they wanted him to hear them.

Mrs. Ralph talked.

"She is the most primeval person I know. She accepts her fate without a trace of resentment; she simply accepts it."

"What is her fate?" asked Lord Walderhurst, still gazing in his unbiassed manner through his monocle, and not turning his head as he spoke.

"It is her fate to be a woman who is perfectly well born, and who is as penniless as a charwoman, and works like one. She is at the beck and call of any one who will give her an odd job to earn a meal with. That is one of the new ways women have found of making a living."

"Good skin," remarked Lord Walderhurst, irrelevantly. "Good hair—quite a lot."

"She has some of the nicest blood in England in her veins, and she engaged my last cook for me," said Mrs. Ralph.

"Hope she was a good cook."

"Very. Emily Fox-Seton has a faculty of finding decent people. I believe it is because she is so decent herself"—with a little laugh.

"Looks quite decent," commented Walderhurst. The knitting was getting on famously.

"It was odd you should see Sir Bruce Norman that day," Agatha Slade was saying. "It must have been just before he was called away to India."

"It was. He sailed the next day. I happen to know, because some friends of mine met me only a few yards from your picture and began to talk about him. I had not known before that he was so rich. I had not heard about his collieries in Lancashire. Oh!"— opening her big eyes in heart-felt yearning,—"how I wish I owned a colliery! It must be so *nice* to be rich!"

"I never was rich," answered Lady Agatha, with a bitter little sigh. "I know it is hideous to be poor."

"*I* never was rich," said Emily, "and I never shall be. You"—a little shyly—"are so different."

Lady Agatha flushed delicately again.

Emily Fox-Seton made a gentle joke. "You have eyes like blue flowers," she said. Lady Agatha lifted the eyes like blue flowers, and they were pathetic.

"Oh!" she gave forth almost impetuously, "sometimes it seems as if it does not matter whether one has eyes or not."

It was a pleasure to Emily Fox-Seton to realise that after this the beauty seemed to be rather drawn toward her. Their acquaintance became almost a sort of intimacy over the wool scarf for the deep-sea fisherman, which was taken up and laid down, and even carried out on the lawn and left under the trees for the footmen to restore when they brought in the rugs and cushions. Lady Maria was amusing herself with the making of knitted scarfs and helmets just now, and bits of white or gray knitting were the fashion at Mallowe. Once Agatha brought hers to Emily's room in the afternoon to ask that a dropped stitch might be taken up, and this established a sort of precedent. Afterward they began to exchange visits.

The strenuousness of things was becoming, in fact, almost too much for Lady Agatha. Most unpleasant things were happening at home, and occasionally Castle Clare loomed up grayly in the distance like a spectre. Certain tradespeople who ought, in Lady Claraway's opinion, to have kept quiet and waited in patience until things became better, were becoming hideously persistent. In view of the fact that Alix's next season must be provided for, it was most awkward. A girl could not be presented and properly launched in the world, in a way which would give her a proper chance, without expenditure. To the Claraways expenditure meant credit, and there were blots as of tears on the letters in which Lady Claraway reiterated that the tradespeople were behaving horribly. Sometimes, she said once in desperation, things looked as if they would all be obliged to shut themselves up in Castle Clare to retrench; and then what was to become of Alix and her season? And there were Millicent and Hilda and Eve.

More than once there was the mist of tears in the flower-blue eyes when Lady Agatha came to talk. Confidence between two women establishes itself through processes at once subtle and simple. Emily Fox-Seton could not have told when she first began to know that the beauty was troubled and distressed; Lady Agatha did not know when she first slipped into making little frank speeches about herself; but these things came about. Agatha found something like comfort in her acquaintance with the big, normal, artless creature—something which actually raised her spirits when she was depressed. Emily Fox-Seton paid constant kindly tribute to her charms, and helped her to believe in them. When she was with her, Agatha always felt that she really was lovely, after all, and that loveliness was a great capital. Emily admired and revered it so, and evidently never dreamed of doubting its omnipotence. She used to talk as if any girl who was a beauty was a potential duchess. In fact, this was a thing she quite ingenuously believed. She had not lived in a world where marriage was a thing of romance, and, for that matter, neither had Agatha. It was nice if a girl liked the man who married her, but if he was a well-behaved, agreeable person, of good means, it was natural that she would end by liking him sufficiently; and to be provided for comfortably or luxuriously for life, and not left upon one's own hands or one's parents', was a thing to be thankful for in any case. It was such a relief to everybody to know that a girl was "settled," and especially it was such a relief to the girl herself. Even novels and plays were no longer fairy-stories of entrancing young men and captivating young women who fell in love with each other in the first chapter, and after increasingly picturesque incidents were married in the last one in the absolute surety of being blissfully happy forevermore. Neither Lady Agatha nor Emily had been brought up on this order of literature, nor in an atmosphere in which it was accepted without reservation.

They had both had hard lives, and knew what lay before them. Agatha knew she must make a marriage or fade out of existence in prosaic and narrowed dulness. Emily knew that there was no prospect for her of desirable marriage at all. She was too poor, too entirely unsupported by social surroundings, and not sufficiently radiant to catch the roving eye. To be able to maintain herself decently, to be given an occasional treat by her more fortunate friends, and to be allowed by fortune to present to the face of the world the appearance of a woman who was not a pauper, was all she could expect. But she felt that Lady Agatha had the right to more. She did not reason the matter out and ask herself why she had the right to more, but she accepted the proposition as a fact. She was ingenuously interested in her fate, and affectionately sympathetic. She used to look at Lord Walderhurst quite anxiously at times when he was talking to the girl. An anxious mother could scarcely have regarded him with a greater desire to analyse his sentiments. The match would be such a fitting one. He would make such an excellent husband—and there were three places, and the diamonds were magnificent. Lady Maria had described to her a certain tiara which she frequently pictured to herself as glittering above Agatha's exquisite low brow. It would be infinitely more becoming to her than to Miss Brooke or Mrs. Ralph, though either of them would have worn it with spirit. She could not help feeling that both Mrs. Ralph's brilliancy and Miss Brooke's insouciant prettiness were not unworthy of being counted in the running, but Lady Agatha seemed somehow so much more completely the thing wanted. She was anxious that she should always look her best, and when she knew that disturbing letters were fretting her, and saw that they made her look pale and less luminous, she tried to raise her spirits.

"Suppose we take a brisk walk," she would say, "and then you might try a little nap. You look a little tired."

"Oh," said Agatha one day, "how kind you are to me! I believe you actually care about my complexion—about my looking well."

"Lord Walderhurst said to me the other day," was Emily's angelically tactful answer, "that you were the only woman he had ever seen who *always* looked lovely."

"Did he?" exclaimed Lady Agatha, and flushed sweetly. "Once Sir Bruce Norman actually said that to me. I told him it was the nicest thing that could be said to a woman. It is all the nicer"—with a sigh—"because it isn't *really* true."

"I am sure Lord Walderhurst believed it true," Emily said. "He is not a man who talks, you know. He is very serious and dignified." She had herself a reverence and admiration for Lord Walderhurst bordering on tender awe. He was indeed a well-mannered person, of whom painful things were not said. He also conducted himself well toward his tenantry, and was patron of several notable charities. To the unexacting and innocently respectful mind of Emily Fox-Seton this was at once impressive and attractive. She knew, though not intimately, many noble personages quite unlike him. She was rather early Victorian and touchingly respectable.

"I have been crying," confessed Lady Agatha.

"I was afraid so, Lady Agatha," said Emily.

"Things are getting hopeless in Curzon Street. I had a letter from Millicent this morning. She is next in age to Alix, and she says—oh, a number of things. When girls see everything passing by them, it makes them irritable. Millicent is seventeen, and she is too lovely. Her hair is like a red-gold cloak, and her eyelashes are twice as long as mine." She sighed again, and her lips, which were like curved rose-petals, unconcealedly quivered. "They were *all* so cross about Sir Bruce Norman going to India," she added.

"He will come back," said Emily, benignly; "but he may be too late. Has he"—ingenuously—"seen Alix?"

Agatha flushed oddly this time. Her delicate skin registered every emotion exquisitely. "He has seen her, but she was in the school-room, and—I don't think—"

She did not finish, but stopped uneasily, and sat and gazed out of the open window into the park. She did not look happy.

The episode of Sir Bruce Norman was brief and even vague. It had begun well. Sir Bruce had met the beauty at a ball, and they had danced together more than once. Sir Bruce had attractions other than his old baronetcy and his coal-mines. He was a good-looking person, with a laughing brown eye and a nice wit. He had danced charmingly and paid gay compliments. He would have done immensely well. Agatha had liked him. Emily sometimes thought she had liked him very much. Her mother had liked him and had thought he was attracted. But after a number of occasions of agreeable meetings, they had encountered each other on the lawn at Goodwood, and he had announced that he was going to India. Forthwith he had gone, and Emily had gathered that somehow Lady Agatha had been considered somewhat to blame. Her people were not vulgar enough to express this frankly, but she had felt it. Her younger sisters had, upon the whole, made her feel it most. It had been borne in upon her that if Alix, or Millicent with the red-gold cloak, or even Eve, who was a gipsy, had been given such a season and such Doucet frocks, they would have combined them with their wonderful complexions and lovely little chins and noses in such a manner as would at least have prevented desirable acquaintances from feeling free to take P. and O. steamers to Bombay.

In her letter of this morning, Millicent's temper had indeed got somewhat the better of her taste and breeding, and lovely Agatha had cried large tears. So it was comforting to be told that Lord Walderhurst had said such an extremely amiable thing. If he was not young, he was really *very* nice, and there were exalted persons who absolutely had rather a fad for him. It would be exceptionally brilliant.

The brisk walk was taken, and Lady Agatha returned from it blooming. She was adorable at dinner, and in the evening gathered an actual court about her. She was all in pink, and a wreath of little pink wild roses lay close about her head, making her, with her tall young slimness, look like a Botticelli nymph. Emily saw that Lord Walderhurst looked at her a great deal. He sat on an extraordinarily comfortable corner seat, and stared through his monocle.

Lady Maria always gave her Emily plenty to do. She had a nice taste in floral arrangement, and early in her visit it had fallen into her hands as a duty to "do" the flowers.

The next morning she was in the gardens early, gathering roses with the dew on them, and was in the act of cutting some adorable "Mrs. Sharman Crawfords," when she found it behoved her to let down her carefully tucked up petticoats, as the Marquis of Walderhurst was walking straight toward her. An instinct told her that he wanted to talk to her about Lady Agatha Slade.

"You get up earlier than Lady Agatha," he remarked, after he had wished her "Good-morning."

"She is oftener invited to the country than I am," she answered. "When I have a country holiday, I want to spend every moment of it out of doors. And the mornings are so lovely. They are not like this in Mortimer Street."

"Do you live in Mortimer Street?"

"Yes."

"Do you like it?"

"I am very comfortable. I am fortunate in having a nice landlady. She and her daughter are very kind to me."

The morning was indeed heavenly. The masses of flowers were drenched with dew, and the already hot sun was drawing fragrance from them and filling the warm air with it. The marquis, with hia monocle fixed, looked up into the cobalt-blue sky and among the trees, where a wood-dove or two cooed with musical softness.

"Yes," he observed, with a glance which swept the scene, "it is different from Mortimer Street, I suppose. Are you fond of the country?"

"Oh, yes," sighed Emily; "oh, yes!"

She was not a specially articulate person. She could not have conveyed in words all that her "Oh, yes!" really meant of simple love for and joy in rural sights and sounds and scents. But when she lifted her big kind hazel eyes to him, the earnestness of her emotion made them pathetic, as the unspeakableness of her pleasures often did.

Lord Walderhurst gazed at her through the monocle with an air he sometimes had of taking her measure without either unkindliness or particular interest.

"Is Lady Agatha fond of the country?" he inquired.

"She is fond of everything that is beautiful," she replied. "Her nature is as lovely as her face, I think."

"Is it?"

Emily walked a step or two away to a rose climbing up the gray-red wall, and began to clip off blossoms, which tumbled sweetly into her basket.

"She seems lovely in everything," she said, "in disposition and manner and—everything. She never seems to disappoint one or make mistakes."

"You are fond of her?"

"She has been so kind to me."

"You often say people are kind to you."

Emily paused and felt a trifle confused. Realising that she was not a clever person, and being a modest one, she began to wonder if she was given to a parrot-phrase which made her tiresome. She blushed up to her ears.

"People are kind," she said hesitatingly. "I—you see, I have nothing to give, and I always seem to be receiving."

"What luck!" remarked his lordship, calmly gazing at her.

He made her feel rather awkward, and she was at once relieved and sorry when he walked away to join another early riser who had come out upon the lawn. For some mysterious reason Emily Fox-Seton liked him. Perhaps his magnificence and the constant talk she had heard of him had warmed her imagination. He had never said anything particularly intelligent to her, but she felt as if he had. He was a rather silent man, but never looked stupid. He had made some good speeches in the House of Lords, not brilliant, but sound and of a dignified respectability. He had also written two pamphlets. Emily had an enormous respect for intellect, and frequently, it must be admitted, for the thing which passed for it. She was not exacting.

During her stay at Mallowe in the summer, Lady Maria always gave a village treat. She had given it for forty years, and it was a lively function. Several hundred wildly joyous village children were fed to repletion with exhilarating buns and cake, and tea in mugs, after which they ran races for prizes, and were entertained in various ways, with the aid of such of the house-party as were benevolently inclined to make themselves useful.

Everybody was not so inclined, though people always thought the thing amusing. Nobody objected to looking on, and some were agreeably stimulated by the general sense of festivity. But Emily Fox-Seton was found by Lady Maria to be invaluable on this occasion. It was so easy, without the least sense of ill-feeling, to give her all the drudgery to do. There was plenty of drudgery, though it did not present itself to Emily Fox-Seton in that light. She no more realised that she was giving Lady Maria a good deal for her money, so to speak, than she realised that her ladyship, though an amusing and delightful, was an absolutely selfish and inconsiderate old woman. So long as Emily Fox-Seton did not seem obviously tired, it would not have occurred to Lady Maria that she could be so; that, after all, her legs and arms were mere human flesh and blood, that her substantial feet were subject to the fatigue unending trudging to and fro induces. Her ladyship was simply delighted that the preparations went so well, that she could turn to Emily for service and always find her ready. Emily made lists and calculations, she worked out plans and made purchases. She interviewed the village matrons who made the cake and buns, and boiled the tea in bags in a copper; she found the women who could be engaged to assist in cutting cake and bread-and-butter and helping to serve it; she ordered the putting up of tents and forms and tables; the innumerable things to be remembered she called to mind.

"Really, Emily," said Lady Maria, "I don't know how I have done this thing for forty years without you. I must always have you at Mallowe for the treat."

Emily was of the genial nature which rejoices upon even small occasions, and is invariably stimulated to pleasure by the festivities of others. The festal atmosphere was a delight to her. In her numberless errands to the village, the sight of the excitement in the faces of the children she passed on her way to this cottage and that filled her eyes with friendly glee and wreathed her face with smiles. When she went into the cottage where the cake was being baked, children hovered about in groups and nudged each other, giggling. They hung about, partly through thrilled interest, and partly because their joy made them eager to courtesy to her as she came out, the obeisance seeming to identify them even more closely with the coming treat. They grinned and beamed rosily, and Emily smiled at them and nodded, uplifted by a pleasure almost as infantile as their own. She was really enjoying herself so honestly that she did not realise how hard she worked during the days before the festivity. She was really ingenious, and invented a number of new methods of entertainment. It was she who, with the aid of a couple of gardeners, transformed the tents into bowers of green boughs and arranged the decorations of the tables and the park gates.

"What a lot of walking you do!" Lord Walderhurst said to her once, as she passed the group on the lawn. "Do you know how many hours you have been on your feet to-day?"

"I like it," she answered, and, as she hurried by, she saw that he was sitting a shade nearer to Lady Agatha than she had ever seen him sit before, and that Agatha, under a large hat of white gauze frills, was looking like a seraph, so sweet and shining were her eyes, so flower-fair her face. She looked actually happy.

"Perhaps he has been saying things," Emily thought. "How happy she will be! He has such a nice pair of eyes. He would make a woman very happy." A faint sigh fluttered from her lips. She was beginning to be physically tired, and was not yet quite aware of it. If she had not been physically tired, she would not even vaguely have had, at this moment, recalled to her mind the fact that she was not of the women to whom "things" are said and to whom things happen.

"Emily Fox-Seton," remarked Lady Maria, fanning herself, as it was frightfully hot, "has the most admirable effect on me. She makes me feel generous. I should like to present her with the smartest things from the wardrobes of all my relations."

"Do you give her clothes?" asked Walderhurst.

"I haven't any to spare. But I know they would be useful to her. The things she wears are touching; they are so well contrived, and produce such a decent effect with so little."

Lord Walderhurst inserted his monocle and gazed after the straight, well-set-up back of the disappearing Miss Fox-Seton.

"I think," said Lady Agatha, gently, "that she is really handsome."

"So she is," admitted Walderhurst—"quite a good-looking woman."

That night Lady Agatha repeated the amiability to Emily, whose grateful amazement really made her blush.

"Lord Walderhurst knows Sir Bruce Norman," said Agatha. "Isn't it strange? He spoke of him to me to-day. He says he is clever."

"You had a nice talk this afternoon, hadn't you?" said Emily. "You both looked so—so—as if you were enjoying yourselves when I passed."

"Did he look as if he were enjoying himself? He was very agreeable. I did not know he could be so agreeable."

"I have never seen him look as much pleased," answered Emily Fox-Seton. "Though he always looks as if he liked talking to you,

Lady Agatha. That large white gauze garden-hat"—reflectively—
"is so *very* becoming."

"It was very expensive," sighed lovely Agatha. "And they last
such a short time. Mamma said it really seemed almost criminal to
buy it."

"How delightful it will be," remarked cheering Emily, "when—
when you need not think of things like that!"

"Oh!"—with another sigh, this time a catch of the breath,—
"it would be like Heaven! People don't know; they think girls
are frivolous when they care, and that it isn't serious. But when
one knows one *must* have things,—that they are like bread,—it is
awful!"

"The things you wear really matter." Emily was bringing all
her powers to bear upon the subject, and with an anxious kindness
which was quite angelic. "Each dress makes you look like another
sort of picture. Have you,"—contemplatively—";anything *quite*
different to wear to-night and to-morrow?"

"I have two evening dresses I have not worn here yet"—a little
hesitatingly. "I—well I saved them. One is a very thin black one
with silver on it. It has a trembling silver butterfly for the shoulder,
and one for the hair."

"Oh, put that on to-night!" said Emily, eagerly. "When you
come down to dinner you will look so—so new! I always think that
to see a fair person suddenly for the first time all in black gives one
a kind of delighted start—though start isn't the word, quite. Do
put it on."

Lady Agatha put it on. Emily Fox-Seton came into her room
to help to add the last touches to her beauty before she went down
to dinner. She suggested that the fair hair should be dressed even
higher and more lightly than usual, so that the silver butterfly should
poise the more airily over the knot, with its quivering, outstretched
wings. She herself poised the butterfly high upon the shoulder.

"Oh, it is lovely!" she exclaimed, drawing back to gaze at the
girl. "Do let me go down a moment or so before you do, so that I
can see you come into the room."

She was sitting in a chair quite near Lord Walderhurst when
her charge entered. She saw him really give something quite like a
start when Agatha appeared. His monocle, which had been in his

eye, fell out of it, and he picked it up by its thin cord and replaced it.

"Psyche!" she heard him say in his odd voice, which seemed merely to make a statement without committing him to an opinion—"Psyche!"

He did not say it to her or to any one else. It was simply a kind of exclamation,—appreciative and perceptive without being enthusiastic,—and it was curious. He talked to Agatha nearly all the evening.

Emily came to Lady Agatha before she retired, looking even a little flushed.

"What are you going to wear at the treat to-morrow?" she asked.

"A white muslin, with *entre-deux* of lace, and the gauze garden-hat, and a white parasol and shoes."

Lady Agatha looked a little nervous; her pink fluttered in her cheek.

"And to-morrow night?" said Emily.

"I have a very pale blue. Won't you sit down, dear Miss Fox-Seton?"

"We must both go to bed and sleep. You must not get tired."

But she sat down for a few minutes, because she saw the girl's eyes asking her to do it.

The afternoon post had brought a more than usually depressing letter from Curzon Street. Lady Claraway was at her motherly wits' ends, and was really quite touching in her distraction. A dressmaker was entering a suit. The thing would get into the papers, of course.

"Unless something happens, something to save us by staving off things, we shall have to go to Castle Clare at once. It will be all over. No girl could be presented with such a thing in the air. They don't like it."

"They," of course, meant persons whose opinions made London's society's law.

"To go to Castle Clare," faltered Agatha, "will be like being sentenced to starve to death. Alix and Hilda and Millicent and Eve and I will be starved, quite slowly, for the want of the things that make girls' lives bearable when they have been born in a certain class. And even if the most splendid thing happened in three or four years, it would be too late for us four—almost too late for Eve. If you are out of London, of course you are forgotten. People can't help forgetting. Why shouldn't they, when there are such crowds of new girls every year?"

Emily Fox-Seton was sweet. She was quite sure that they would not be obliged to go to Castle Clare. Without being indelicate, she was really able to bring hope to the fore. She said a good deal of the black gauze dress and the lovely effect of the silver butterflies.

"I suppose it was the butterflies which made Lord Walderhurst say 'Psyche! Psyche!' when he first saw you," she added, *en passant*.

"Did he say that?" And immediately Lady Agatha looked as if she had not intended to say the words.

"Yes," answered Emily, hurrying on with a casual air which had a good deal of tact in it. "And black makes you so wonderfully fair and aërial. You scarcely look quite real in it; you might float away. But you must go to sleep now."

Lady Agatha went with her to the door of the room to bid her good-night. Her eyes looked like those of a child who might presently cry a little. "Oh, Miss Fox-Seton," she said, in a very young voice, "you are so kind!"

CHAPTER FOUR

The parts of the park nearest to the house already presented a busy aspect when Miss Fox-Seton passed through the gardens the following morning. Tables were being put up, and baskets of bread and cake and groceries were being carried into the tent where the tea was to be prepared. The workers looked interested and good-humoured; the men touched their hats as Emily appeared, and the women courtesied smilingly. They had all discovered that she was amiable and to be relied on in her capacity of her ladyship's representative.

"She's a worker, that Miss Fox-Seton," one said to the other. "I never seen one that was a lady fall to as she does. Ladies, even when they means well, has a way of standing about and telling you to do things without seeming to know quite how they ought to be done. She's coming to help with the bread-and-butter-cutting herself this morning, and she put up all them packages of sweets yesterday with her own hands. She did 'em up in different-coloured papers, and tied 'em with bits of ribbon, because she said she knowed children was prouder of coloured things than plain—they was like that. And so they are: a bit of red or blue goes a long way with a child."

Emily cut bread-and-butter and cake, and placed seats and arranged toys on tables all the morning. The day was hot, though beautiful, and she was so busy that she had scarcely time for her breakfast. The household party was in the gayest spirits. Lady Maria was in her most amusing mood. She had planned a drive to some interesting ruins for the afternoon of the next day, and a dinner-party for the evening. Her favourite neighbours had just returned to their country-seat five miles away, and they were coming to the dinner, to her great satisfaction. Most of her neighbours bored her, and she took them in doses at her dinners, as she would have taken medicine. But the Lockyers were young and good-looking and clever, and she was always glad when they came to Loche during her stay at Mallowe.

"There is not a frump or a bore among them," she said. "In the country people are usually frumps when they are not bores, and bores when they are not frumps, and I am in danger of becoming both myself. Six weeks of unalloyed dinner-parties, composed of certain people I know, would make me begin to wear moreen petticoats and talk about the deplorable condition of London society."

She led all her flock out on to the lawn under the ilex-trees after breakfast.

"Let us go and encourage industry," she said. "We will watch Emily Fox-Seton working. She is an example."

Curiously enough, this was Miss Cora Brooke's day. She found herself actually walking across the lawn with Lord Walderhurst by her side. She did not know how it happened, but it seemed to occur accidentally.

"We never talk to each other," he said.

"Well," answered Cora, "we have talked to other people a great deal—at least I have."

"Yes, you have talked a good deal," said the marquis.

"Does that mean I have talked too much?"

He surveyed her prettiness through his glass. Perhaps the holiday stir in the air gave him a festive moment.

"It means that you haven't talked enough to me. You have devoted yourself too much to the laying low of young Heriot."

She laughed a trifle saucily.

"You are a very independent young lady," remarked Walderhurst, with a lighter manner than usual. "You ought to say something deprecatory or—a little coy, perhaps."

"I shan't," said Cora, composedly.

"Shan't or won't?" he inquired. "They are both bad words for little girls—or young ladies—to use to their elders."

"Both," said Miss Cora Brooke, with a slightly pleased flush. "Let us go over to the tents and see what poor Emily Fox-Seton is doing."

"Poor Emily Fox-Seton," said the marquis, non-committally.

They went, but they did not stay long. The treat was taking form. Emily Fox-Seton was hot and deeply engaged. People were coming to her for orders. She had a thousand things to do and to superintend the doing of. The prizes for the races and the presents for the children must be arranged in order: things for boys and things for girls, presents for little children and presents for big ones. Nobody must be missed, and no one must be given the wrong thing.

"It would be dreadful, you know," Emily said to the two when they came into her tent and began to ask questions, "if a big boy should get a small wooden horse, or a little baby should be given a cricket bat and ball. Then it would be so disappointing if a tiny girl got a work-box and a big one got a doll. One has to get things in order. They look forward to this so, and it's heart-breaking to a child to be disappointed, isn't it?"

Walderhurst gazed uninspiringly.

"Who did this for Lady Maria when you were not here?" he inquired.

"Oh, other people. But she says it was tiresome." Then with an illumined smile; "She has asked me to Mallowe for the next twenty years for the treats. She is so kind."

"Maria is a kind woman"—with what seemed to Emily delightful amiability. "She is kind to her treats and she is kind to Maria Bayne."

"She is kind to *me*," said Emily. "You don't know how I am enjoying this."

"That woman enjoys everything," Lord Walderhurst said when he walked away with Cora. "What a temperament to have! I would give ten thousand a year for it."

"She has so little," said Cora, "that everything seems beautiful to her. One doesn't wonder, either. She's very nice. Mother and I quite admire her. We are thinking of inviting her to New York and giving her a real good time."

"She would enjoy New York."

"Have you ever been there, Lord Walderhurst?"

"No."

"You ought to come, really. So many Englishmen come now, and they all seem to like it."

"Perhaps I will come," said Walderhurst. "I have been thinking of it. One is tired of the Continent and one knows India. One doesn't know Fifth Avenue, and Central Park, and the Rocky Mountains."

"One might try them," suggested pretty Miss Cora.

This certainly was her day. Lord Walderhurst took her and her mother out in his own particular high phaeton before lunch. He was fond of driving, and his own phaeton and horses had come to Mallowe with him. He took only his favourites out, and though he bore himself on this occasion with a calm air, the event caused a little smiling flurry on the lawn. At least, when the phaeton spun down the avenue with Miss Brooke and her mother looking slightly flushed and thrilled in their high seats of honour, several people exchanged glances and raised eye-brows.

Lady Agatha went to her room and wrote a long letter to Curzon Street. Mrs. Ralph talked about the problem-play to young Heriot and a group of others.

The afternoon, brilliant and blazing, brought new visitors to assist by their presence at the treat. Lady Maria always had a large house-party, and added guests from the neighbourhood to make for gaiety. At two o'clock a procession of village children and their friends and parents, headed by the village band, marched up the avenue and passed before the house on their way to their special part of the park. Lady Maria and her guests stood upon the broad steps and welcomed the jocund crowd, as it moved by, with hospitable bows and nods and becks and wreathed smiles. Everybody was in a delighted good-humour.

As the villagers gathered in the park, the house-party joined them by way of the gardens. A conjurer from London gave an entertainment under a huge tree, and children found white rabbits taken from their pockets and oranges from their caps, with squeals of joy and shouts of laughter. Lady Maria's guests walked about and looked on, laughing with the children.

The great affair of tea followed the performance. No treat is fairly under way until the children are filled to the brim with tea and buns and cake, principally cake in plummy wedges.

Lady Agatha and Mrs. Ralph handed cake along rows of children seated on the grass. Miss Brooke was talking to Lord Walderhurst when the work began. She had poppies in her hat and carried a poppy-coloured parasol, and sat under a tree, looking very alluring.

"I ought to go and help to hand cake," she said.

"My cousin Maria ought to do it," remarked Lord Walderhurst, "but she will not—neither shall I. Tell me something about the elevated railroad and Five-Hundred-and-Fifty-Thousandth Street." He had a slightly rude, gracefully languid air, which Cora Brooke found somewhat impressive, after all.

Emily Fox-Seton handed cake and regulated supplies with cheerful tact and good spirits. When the older people were given their tea, she moved about their tables, attending to every one. She was too heart-whole in her interest in her hospitalities to find time to join Lady Maria and her party at the table under the ilex-trees. She ate some bread-and-butter and drank a cup of tea while she talked to some old women she had made friends with. She was really enjoying herself immensely, though occasionally she was obliged to sit down for a few moments just to rest her tired feet. The children came to her as to an omnipotent and benign being. She knew where the toys were kept and what prizes were to be given for the races. She represented law and order and bestowal. The other ladies walked about in wonderful dresses, smiling and exalted, the gentlemen aided the sports in an amateurish way and made patrician jokes among themselves, but this one lady seemed to be part of the treat itself. She was not so grandly dressed as the others,—her dress was only blue linen with white bands on it,—and she had only a sailor hat with a buckle and bow, but she was of her ladyship's world of London people, nevertheless, and they liked her more than they had ever liked a lady before. It was a fine treat, and she seemed to have made it so. There had never been quite such a varied and jovial treat at Mallowe before.

The afternoon waxed and waned. The children played games and raced and rejoiced until their young limbs began to fail them. The older people sauntered about or sat in groups to talk and listen to the village band. Lady Maria's visitors, having had enough of rural festivities, went back to the gardens in excellent spirits, to talk and to watch a game of tennis which had taken form on the court.

Emily Fox-Seton's pleasure had not abated, but her colour had done so. Her limbs ached and her still-smiling face was pale, as she stood under the beech-tree regarding the final ceremonies of the festal day, to preside over which Lady Maria and her party returned from their seats under the ilex-trees. The National Anthem was sung loudly, and there were three tremendous cheers given for her ladyship. They were such joyous and hearty cheers that Emily was stirred almost to emotional tears. At all events, her hazel eyes looked nice and moistly bright. She was an easily moved creature.

Lord Walderhurst stood near Lady Maria and looked pleased also. Emily saw him speak to her ladyship and saw Lady Maria smile. Then he stepped forward, with his non-committal air and his monocle glaring calmly in his eye.

"Boys and girls," he said in a clear, far-reaching voice, "I want you to give three of the biggest cheers you are capable of for the lady who has worked to make your treat the success it has been. Her ladyship tells me she has never had such a treat before. Three cheers for Miss Fox-Seton."

Emily gave a gasp and felt a lump rise in her throat. She felt as if she had been without warning suddenly changed into a royal personage, and she scarcely knew what to do.

The whole treat, juvenile and adult, male and female, burst into three cheers which were roars and bellows. Hats and caps were waved and tossed into the air, and every creature turned toward her as she blushed and bowed in tremulous gratitude and delight.

"Oh, Lady Maria! oh, Lord Walderhurst!" she said, when she managed to get to them, "how *kind* you are to me!"

CHAPTER FIVE

After she had taken her early tea in the morning, Emily Fox-Seton lay upon her pillows and gazed out upon the tree-branches near her window, in a state of bliss. She was tired, but happy. How well everything had "gone off"! How pleased Lady Maria had been, and how kind of Lord Walderhurst to ask the villagers to give three cheers for herself! She had never dreamed of such a thing. It was the kind of attention not usually offered to her. She smiled her childlike smile and blushed at the memory of it. Her impression of the world was that people were really very amiable, as a rule. They were always good to her, at least, she thought, and it did not occur to her that if she had not paid her way so remarkably well by being useful they might have been less agreeable. Never once had she doubted that Lady Maria was the most admirable and generous of human beings. She was not aware in the least that her ladyship got a good deal out of her. In justice to her ladyship, it may be said that she was not wholly aware of it herself, and that Emily absolutely enjoyed being made use of.

This morning, however, when she got up, she found herself more tired than she ever remembered being before, and it may be easily argued that a woman who runs about London on other people's errands often knows what it is to be aware of aching limbs. She laughed a little when she discovered that her feet were actually rather swollen, and that she must wear a pair of her easiest slippers. "I must sit down as much as I can to-day," she thought. "And yet, with the dinner-party and the excursion this morning, there may be a number of little things Lady Maria would like me to do."

There were, indeed, numbers of things Lady Maria was extremely glad to ask her to do. The drive to the ruins was to be made before lunch, because some of the guests felt that an afternoon jaunt would leave them rather fagged for the dinner-party in the evening. Lady Maria was not going, and, as presently became apparent, the carriages would be rather crowded if Miss Fox-Seton joined the party. On the whole, Emily was not sorry to have an excuse for remaining at home, and so the carriages drove away comfortably filled, and Lady Maria and Miss Fox-Seton watched their departure.

"I have no intention of having my venerable bones rattled over hill and dale the day I give a dinner-party," said her ladyship. "Please ring the bell, Emily. I want to make sure of the fish. Fish is one of the problems of country life. Fishmongers are demons, and when they live five miles from one they can arouse the most powerful human emotions."

Mallowe Court was at a distance from the country town delightful in its effects upon the rusticity of the neighbourhood, but appalling when considered in connection with fish. One could not dine Without fish; the town was small and barren of resources, and the one fishmonger of weak mind and unreliable nature.

The footman who obeyed the summons of the bell informed her ladyship that the cook was rather anxious about the fish, as usual. The fishmonger had been a little doubtful as to whether he could supply her needs, and his cart never arrived until half-past twelve.

"Great goodness!" exclaimed her ladyship when the man retired. "What a situation if we found ourselves without fish! Old General Barnes is the most ferocious old gourmand in England, and he loathes people who give him bad dinners. We are all rather afraid of him, the fact is, and I will own that I am vain about my dinners. That is the last charm nature leaves a woman, the power to give decent dinners. I shall be fearfully annoyed if any ridiculous thing happens."

They sat in the morning-room together writing notes and talking, and as half-past twelve drew near, watching for the fishmonger's cart. Once or twice Lady Maria spoke of Lord Walderhurst.

"He is an interesting creature, to my mind," she said. "I have always rather liked him. He has original ideas, though he is not in the least brilliant. I believe he talks more freely to me, on the whole, than to most people, though I can't say he has a particularly good opinion of me. He stuck his glass in his eye and stared at me last night, in that weird way of his, and said to me, 'Maria, in an ingenuous fashion of your own, you are the most abominably selfish woman I ever beheld.' Still, I know he rather likes me. I said to him: 'That isn't quite true, James. I am selfish, but I'm not *abominably* selfish. Abominably selfish people always have nasty tempers, and no one can accuse me of having a nasty temper. I have the disposition of a bowl of bread and milk."

"Emily,"—as wheels rattled up the avenue,—"*is* that the fishmonger's cart?"

"No," answered Emily at the window; "it is the butcher."

"His attitude toward the women here has made my joy," Lady Maria proceeded, smiling over the deep-sea fishermen's knitted helmet she had taken up. "He behaves beautifully to them all, but not one of them has really a leg to stand on as far as he is responsible for it. But I will tell you something, Emily." She paused.

Miss Fox-Seton waited with interested eyes.

"He is thinking of bringing the thing to an end and marrying *some* woman. I feel it in my bones."

"Do you think so?" exclaimed Emily. "Oh, I can't help hoping—" But she paused also.

"You hope it will be Agatha Slade," Lady Maria ended for her. "Well, perhaps it will be. I sometimes think it is Agatha, if it's any one. And yet I'm not sure. One never could be sure with Walderhurst. He has always had a trick of keeping more than his mouth shut. I wonder if he could have any other woman up his sleeve?"

"Why do you think—" began Emily.

Lady Maria laughed.

"For an odd reason. The Walderhursts have a ridiculously splendid ring in the family, which they have a way of giving to the women they become engaged to. It's ridiculous because—well, because a ruby as big as a trouser's button *is* ridiculous. You can't get over that. There is a story connected with this one—centuries and things, and something about the woman the first Walderhurst had it made for. She was a Dame Something or Other who had snubbed the King for being forward, and the snubbing was so good for him that he thought she was a saint and gave the ruby for her betrothal. Well, by the merest accident I found Walderhurst had sent his man to town for it. It came two days ago."

"Oh, how interesting!" said Emily, thrilled. "It *must* mean something."

"It is rather a joke. Wheels again, Emily. Is *that* the fishmonger?"

Emily went to the window once more. "Yes," she answered, "if his name is Buggle."

"His name *is* Buggle," said Lady Maria, "and we are saved."

But five minutes later the cook herself appeared at the morning-room door. She was a stout person, who panted, and respectfully removed beads of perspiration from her brow with a clean handkerchief.

She was as nearly pale as a heated person of her weight may be.

"And what has happened now, cook?" asked Lady Maria.

"That Buggle, your ladyship," said cook, "says your ladyship can't be no sorrier than he is, but when fish goes bad in a night it can't be made fresh in the morning. He brought it that I might see it for myself, and it is in a state as could not be used by any one. I was that upset, your ladyship, that I felt like I must come and explain myself."

"What *can* be done?" exclaimed Lady Maria. "Emily, *do* suggest something."

"We can't even be sure," said the cook, "that Batch has what would suit us. Batch sometimes has it, but he is the fishmonger at Maundell, and that is four miles away, and we are short-'anded, your ladyship, now the 'ouse is so full, and not a servant that could be spared."

"Dear me!" said Lady Maria. "Emily, this is really enough to drive one quite mad. If everything was not out of the stables, I know you would drive over to Maundell. You are such a good walker,"—catching a gleam of hope,—"do you think you could walk?"

Emily tried to look cheerful. Lady Maria's situation was really an awful one for a hostess. It would not have mattered in the least if her strong, healthy body had not been so tired. She was an excellent walker, and ordinarily eight miles would have meant nothing in the way of fatigue. She was kept in good training by her walking in town, Springy moorland swept by fresh breezes was not like London streets.

"I think I can manage it," she said nice-temperedly. "If I had not run about so much yesterday it would be a mere nothing. You must have the fish, of course. I will walk over the moor to Maundell and tell Batch it must be sent at once. Then I will come back slowly. I can rest on the heather by the way. The moor is lovely in the afternoon."

"You dear soul!" Lady Maria broke forth. "What a boon you are to a woman!"

She felt quite grateful. There arose in her mind an impulse to invite Emily Fox-Seton to remain the rest of her life with her, but she was too experienced an elderly lady to give way to impulses. She privately resolved, however, that she would have her a good deal in South Audley Street, and would make her some decent presents.

When Emily Fox-Seton, attired for her walk in her shortest brown linen frock and shadiest hat, passed through the hall, the post-boy was just delivering the midday letters to a footman. The servant presented his salver to her with a letter for herself lying upon the top of one addressed in Lady Claraway's handwriting "To the Lady Agatha Slade." Emily recognised it as one of the epistles of many sheets which so often made poor Agatha shed slow and depressed tears. Her own letter was directed in the well-known hand of Mrs. Cupp, and she wondered what it could contain.

"I hope the poor things are not in any trouble," she thought. "They were afraid the young man in the sitting-room was engaged. If he got married and left them, I don't know what they would do; he has been so regular."

Though the day was hot, the weather was perfect, and Emily, having exchanged her easy slippers for an almost equally easy pair of tan shoes, found her tired feet might still be used. Her disposition to make the very best of things inspired her to regard even an eight-mile walk with courage. The moorland air was so sweet, the sound of the bees droning as they stumbled about in the heather was such a comfortable, peaceful thing, that she convinced herself that she should find the four miles to Maundell quite agreeable.

She had so many nice things to think of that she temporarily forgot that she had put Mrs. Cupp's letter in her pocket, and was half-way across the moor before she remembered it.

"Dear me!" she exclaimed when she recalled it. "I must see what has happened."

She opened the envelope and began to read as she walked; but she had not taken many steps before she uttered an exclamation and stopped.

"How very nice for them!" she said, but she turned rather pale.

From a worldly point of view the news the letter contained was indeed very nice for the Cupps, but it put a painful aspect upon the simple affairs of poor Miss Fox-Seton.

"It is a great piece of news, in one way," wrote Mrs. Cupp, "and yet me and Jane can't help feeling a bit low at the thought of the changes it will make, and us living where you won't be with us, if I may take the liberty, miss. My brother William made a good bit of money in Australia, but he has always been homesick for the old country, as he always calls England. His wife was a Colonial, and when she died a year ago he made up his mind to come home to settle in Chichester, where he was born. He says there's nothing like the feeling of a Cathedral town. He's bought such a nice house a bit out, with a big garden, and he wants me and Jane to come and make a home with him. He says he has worked hard all his life, and now he means to be comfortable, and he can't be bothered with housekeeping. He promises to provide well for us both, and he wants us to sell up Mortimer Street, and come as quick as possible. But we *shall* miss you, miss, and though her Uncle William keeps a trap and everything according, and Jane is grateful for his kindness, she broke down and cried hard last night, and says to me: 'Oh, mother, if Miss Fox-Seton could just manage to take me as a maid, I would rather be it than anything. Traps don't feed the heart, mother, and I've a feeling for Miss Fox-Seton as is perhaps unbecoming to my station.' But we've got the men in the house ticketing things, miss, and we want to know what we shall do with the articles in your bed-sitting-room."

The friendliness of the two faithful Cupps and the humble Turkey-red comforts of the bed-sitting-room had meant home to Emily Fox-Seton. When she had turned her face and her tired feet away from discouraging errands and small humiliations and discomforts, she had turned them toward the bed-sitting-room, the hot little fire, the small, fat black kettle singing on the hob, and the two-and-eleven-penny tea-set. Not being given to crossing bridges before she reached them, she had never contemplated the dreary possibility that her refuge might be taken away from her. She had not dwelt upon the fact that she had no other real refuge on earth.

As she walked among the sun-heated heather and the luxuriously droning bees, she dwelt upon it now with a suddenly realising sense. As it came home to her soul, her eyes filled with big tears, which brimmed over and rolled down her cheeks. They dropped upon the breast of her linen blouse and left marks.

"I shall have to find a new bed-sitting-room somewhere," she said, the breast of the linen blouse lifting itself sharply. "It will be so different to be in a house with strangers. Mrs. Cupp and Jane—" She was obliged to take out her handkerchief at that moment. "I am afraid I can't get anything respectable for ten shillings a week. It Was very cheap—and they were so nice!"

All her fatigue of the early morning had returned. Her feet began to burn and ache, and the sun felt almost unbearably hot. The mist in her eyes prevented her seeing the path before her. Once or twice she stumbled over something.

"It seems as if it must be farther than four miles," she said. "And then there is the walk back. I *am* tired. But I must get on, really."

CHAPTER SIX

The drive to the ruins had been a great success. It was a drive of just sufficient length to put people in spirits without fatiguing them. The party came back to lunch with delightful appetites. Lady Agatha and Miss Cora Brooke had pink cheeks. The Marquis of Walderhurst had behaved charmingly to both of them. He had helped each of them to climb about among the ruins, and had taken them both up the steep, dark stairway of one of the towers, and stood with them looking over the turrets into the courtyard and the moat. He knew the history of the castle and could point out the banquet-hall and the chapel and the serving-places, and knew legends about the dungeons.

"He gives us all a turn, mother," said Miss Cora Brooke. "He even gave a turn yesterday to poor Emily Fox-Seton. He's rather nice."

There was a great deal of laughter at lunch after their return. Miss Cora Brooke was quite brilliant in her gay little sallies. But though she was more talkative than Lady Agatha, she did not look more brilliant.

The letter from Curzon Street had not made the beauty shed tears. Her face had fallen when it had been handed to her on her return, and she had taken it upstairs to her room with rather a flagging step. But when she came down to lunch she walked with the movement of a nymph. Her lovely little face wore a sort of tremulous radiance. She laughed like a child at every amusing thing that was said. She might have been ten years old instead of twenty-two, her colour, her eyes, her spirits seemed of a freshness so infantine.

She was leaning back in her chair laughing enchantingly at one of Miss Brooke's sparkling remarks when Lord Walderhurst, who sat next to her, said suddenly, glancing round the table:

"But where is Miss Fox-Seton?"

It was perhaps a significant fact that up to this moment nobody had observed her absence. It was Lady Maria who replied.

"I am almost ashamed to answer," she said. "As I have said before, Emily Fox-Seton has become the lodestar of my existence. I cannot live without her. She has walked over to Maundell to make sure that we do not have a dinner-party without fish to-night."

"She has *walked* over to Maundell," said Lord Walderhurst—"after yesterday?"

"There was not a pair of wheels left in the stable," answered Lady Maria. "It is disgraceful, of course, but she is a splendid walker, and she said she was not too tired to do it. It is the kind of thing she ought to be given the Victoria Cross for—saving one from a dinner-party without fish."

The Marquis of Walderhurst took up the cord of his monocle and fixed the glass rigidly in his eye.

"It is not only four miles to Maundell," he remarked, staring at the table-cloth, not at Lady Maria, "but it is four miles back."

"By a singular coincidence," said Lady Maria.

The talk and laughter went on, and the lunch also, but Lord Walderhurst, for some reason best known to himself, did not finish his. For a few seconds he stared at the table-cloth, then he pushed aside his nearly disposed-of cutlet, then he got up from his chair quietly.

"Excuse me, Maria," he said, and without further ado went out of the room, and walked toward the stables.

There was excellent fish at Maundell; Batch produced it at once, fresh, sound, and desirable. Had she been in heir normal spirits, Emily would have rejoiced at the sight of it, and have retraced her four miles to Mallowe in absolute jubilation. She would have shortened and beguiled her return journey by depicting to herself Lady Maria's pleasure and relief.

But the letter from Mrs. Cupp lay like a weight of lead in her pocket. It had given her such things to think of as she walked that she had been oblivious to heather and bees and fleece-bedecked summer-blue sky, and had felt more tired than in any tramp through London streets that she could call to mind. Each step she took seemed to be carrying her farther away from the few square yards of home the bed-sitting-room had represented under the dominion of the Cupps. Every moment she recalled more strongly that it had been home—home. Of course it had not been the third-floor back room so much as it had been the Cupps who made it so, who had regarded her as a sort of possession, who had liked to serve her, and had done it with actual affection.

"I shall have to find a new place," she kept saying. "I shall have to go among quite strange people."

She had suddenly a new sense of being without resource. That was one of the proofs of the curious heaviness of the blow the simple occurrence was to her. She felt temporarily almost as if there were no other lodging-houses in London, though she knew that really there were tens of thousands. The fact was that though there might be other Cupps, or their counterparts, she could not make herself believe such a good thing possible. She had been physically worn out before she had read the letter, and its effect had been proportionate to her fatigue and lack of power to rebound. She was vaguely surprised to feel that the tears kept filling her eyes and falling on her cheeks in big heavy drops. She was obliged to use her handkerchief frequently, as if she was suddenly developing a cold in her head.

"I must take care," she said once, quite prosaically, but with more pathos in her voice than she was aware of, "or I shall make my nose quite red."

The Marquis of Walderhurst

Though Batch was able to supply fish, he was unfortunately not able to send it to Mallowe. His cart had gone out on a round just before Miss Fox-Seton's arrival, and there was no knowing when it would return.

"Then I must carry the fish myself," said Emily. "You can put it in a neat basket."

"I'm very sorry, miss; I am, indeed, miss," said Batch, looking hot and pained.

"It will not be heavy," returned Emily; "and her ladyship must be sure of it for the dinner-party."

So she turned back to recross the moor with a basket of fish on her arm. And she was so pathetically unhappy that she felt that so long as she lived the odour of fresh fish would make her feel sorrowful. She had heard of people who were made sorrowful by the odour of a flower or the sound of a melody but in her case it would be the smell of fresh fish that would make her sad. If she had been a person with a sense of humour, she might have seen that this was thing to laugh at a little. But she was not a humorous woman, and just now—

"Oh, I shall have to find a new place," she was thinking, "and I have lived in that little room for years."

The sun got hotter and hotter, and her feet became so tired that she could scarcely drag one of them after another. She had forgotten that she had left Mallowe before lunch, and that she ought to have got a cup of tea, at least, at Maundell. Before she had walked a mile on her way back, she realised that she was frightfully hungry and rather faint.

"There is not even a cottage where I could get a glass of water," she thought.

The basket, which was really comparatively light, began to feel heavy on her arm, and at length she felt sure that a certain burning spot on her left heel must be a blister which was being rubbed by her shoe. How it hurt her, and how tired she was—how tired! And when she left Mallowe—lovely, luxurious Mallowe—she would not go back to her little room all fresh from the Cupps' autumn house-cleaning, which included the washing and ironing of her Turkey-red hangings and chair-covers; she would be obliged to huddle into any poor place she could find. And Mrs. Cupp and Jane would be in Chichester.

"But what good fortune it is for them!" she murmured. "They need never be anxious about the future again. How—how wonderful it must be to know that one need not be afraid of the future! I—indeed, I think I really must sit down."

She sat down upon the sun-warmed heather and actually let her tear-wet face drop upon her hands.

"Oh, dear! Oh, dear! Oh, dear!" she said helplessly. "I must not let myself do this. I mustn't, Oh, dear! Oh, dear! Oh, dear!"

She was so overpowered by her sense of her own weakness that she was conscious of nothing but the fact that she must control it. Upon the elastic moorland road wheels stole upon one without sound. So the wheels of a rapidly driven high cart approached her and were almost at her side before she lifted her head, startled by a sudden consciousness that a vehicle was near her.

It was Lord Walderhurst's cart, and even as she gazed at him with alarmed wet eyes, his lordship descended from it and made a sign to his groom, who at once impassively drove on.

Emily's lips tried to tremble into a smile; she put out her hand fumblingly toward the fish-basket, and having secured it, began to rise.

"I—sat down to rest," she faltered, even apologetically. "I walked to Maundell, and it was so hot."

Just at that moment a little breeze sprang up and swept across her cheek. She was so grateful that her smile became less difficult.

"I got what Lady Maria wanted," she added, and the childlike dimple in her cheek endeavoured to defy her eyes.

The Marquis of Walderhurst looked rather odd. Emily had never seen him look like this before. He took a silver flask out of his pocket in a matter-of-fact way, and filled its cup with something.

"That is sherry," he said. "Please drink it. You are absolutely faint."

She held out her hand eagerly. She could not help it.

"Oh, thank you—thank you!" she said. "I am *so* thirsty!" And she drank it as if it were the nectar of the gods.

"Now, Miss Fox-Seton," he said, "please sit down again. I came here to drive you back to Mallowe, and the cart will not come back for a quarter of an hour."

"You came on purpose!" she exclaimed, feeling, in truth, somewhat awe-struck. "But how kind of you, Lord Walderhurst—how good!"

It was the most unforeseen and amazing experience of her life, and at once she sought for some reason which could connect with his coming some more interesting person than mere Emily Fox-Seton. Oh,—the thought flashed upon her,—he had come for some reason connected with Lady Agatha. He made her sit down on the heather again, and he took a seat beside her. He looked straight into her eyes.

"You have been crying," he remarked.

There was no use denying it. And what was there in the good gray-brown eye, gazing through the monocle, which so moved her by its suggestion of kindness and—and some new feeling?

"Yes, I have," she admitted. "I don't often—but—well, yes, I have."

"What was it?"

It was the most extraordinary thump her heart gave at this moment. She had never felt such an absolute thump. It was perhaps because she was tired. His voice had lowered itself. No man had ever spoken to her before like that. It made one feel as if he was not an exalted person at all; only a kind, kind one. She must not presume upon his kindness and make much of her prosaic troubles. She tried to smile in a proper casual way.

"Oh, it was a small thing, really," was her effort at treating the matter lightly; "but it seems more important to me than it would to any one with—with a family. The people I live with—who have been so kind to me—are going away."

"The Cupps?" he asked.

She turned quite round to look at him.

"How," she faltered, "did you know about them?"

"Maria told me," he answered, "I asked her."

It seemed such a human sort of interest to have taken in her. She could not understand. And she had thought he scarcely realised her existence. She said to herself that was so often the case—people were so much kinder than one knew.

She felt the moisture welling in her eyes, and stared steadily at the heather, trying to wink it away.

"I am really glad," she explained hastily. "It is such good fortune for them. Mrs. Cupp's brother has offered them such a nice home. They need never be anxious again."

"But they will leave Mortimer Street—and you will have to give up your room."

"Yes. I must find another." A big drop got the better of her, and flashed on its way down her cheek. "I can find a room, perhaps, but—I can't find—" She was obliged to clear her throat.

"That was why you cried?"

"Yes." After which she sat still.

"You don't know where you will live?"

"No."

She was looking so straight before her and trying so hard to behave discreetly that she did not see that he had drawn nearer to her. But a moment later she realised it, because he took hold of her hand. His own closed over it firmly.

"Will you," he said—"I came here, in fact, to ask you if you will come and live with me?"

Her heart stood still, quite still. London was so full of ugly stories about things done by men of his rank—stories of transgressions, of follies, of cruelties. So many were open secrets. There were men, who, even while keeping up an outward aspect of respectability, were held accountable for painful things. The lives of well-born struggling women were so hard. Sometimes such nice ones went under because temptation was so great. But she had not thought, she could not have dreamed—

She got on her feet and stood upright before him. He rose with her, and because she was a tall woman their eyes were on a level. Her own big and honest ones were wide and full of crystal tears.

"Oh!" she said in helpless woe. "Oh!"

It was perhaps the most effective thing a woman ever did. It was so simple that it was heartbreaking. She could not have uttered a word, he was such a powerful and great person, and she was so without help or stay.

Since the occurring of this incident, she has often been spoken of as a beauty, and she has, without doubt, had her fine hours; but Walderhurst has never told her that the most beautiful moment of her life was undoubtedly that in which she stood upon the heather, tall and straight and simple, her hands hanging by her sides, her large, tear-filled hazel eyes gazing straight into his. In the femininity of her frank defencelessness there was an appeal to nature's self in man which was not quite of earth. And for several seconds they stood so and gazed into each other's souls—the usually unilluminated nobleman and the prosaic young woman who lodged on a third floor back in Mortimer Street.

Then, quite quickly, something was lighted in his eyes, and he took a step toward her.

"Good heavens!" he demanded. "What do you suppose I am asking of you?"

"I don't—know," she answered; "I don't—know."

"My good girl," he said, even with some irritation, "I am asking you to be my wife. I am asking you to come and live with me in an entirely respectable manner, as the Marchioness of Walderhurst."

Emily touched the breast of her brown linen blouse with the tips of her fingers.

"You—are—asking—*me?*" she said.

"Yes," he answered. His glass had dropped out of his eye, and he picked it up and replaced it. "There is Black with the cart," he said. "I will explain myself with greater clearness as we drive back to Mallowe."

The basket of fish was put in the cart, and Emily Fox-Seton was put in. Then the marquis got in himself, and took the reins from his groom.

"You will walk back, Black," he said, "by that path," with a wave of the hand in a diverging direction.

As they drove across the heather, Emily was trembling softly from head to foot. She could have told no human being what she felt. Only a woman who had lived as she had lived and who had been trained as she had been trained could have felt it. The brilliance of the thing which had happened to her was so unheard of and so undeserved, she told herself. It was so incredible that, even with the splendid gray mare's high-held head before her and Lord Walderhurst by her side, she felt that she was only part of a dream. Men had never said "things" to her, and a man was saying them—the Marquis of Walderhurst was saying them. They were not the kind of things every man says or said in every man's way, but they so moved her soul that she quaked with joy.

"I am not a marrying man," said his lordship, "but I must marry, and I like you better than any woman I have ever known. I do not generally like women. I am a selfish man, and I want an unselfish woman. Most women are as selfish as I am myself. I used to like you when I heard Maria speak of you. I have watched you and thought of you ever since I came here. You are necessary to every one, and you are so modest that you know nothing about it. You are a handsome woman, and you are always thinking of other women's good looks."

Emily gave a soft little gasp.

"But Lady Agatha," she said. "I was sure it was Lady Agatha."

"I don't want a girl," returned his lordship. "A girl would bore me to death. I am not going to dry-nurse a girl at the age of fifty-four. I want a companion."

"But I am so *far* from clever," faltered Emily.

The marquis turned in his driving-seat to look at her. It was really a very nice look he gave her. It made Emily's cheeks grow pink and her simple heart beat.

"You are the woman I want," he said. "You make me feel quite sentimental."

When they reached Mallowe, Emily had upon her finger the ruby which Lady Maria had graphically described as being "as big as a trouser button." It was, indeed, so big that she could scarcely wear her glove over it. She was still incredible, but she was blooming like a large rose. Lord Walderhurst had said so many "things" to her that she seemed to behold a new heaven and a new earth. She had been so swept off her feet that she had not really been allowed time to think, after that first gasp, of Lady Agatha.

When she reached her bedroom she almost returned to earth as she remembered it. Neither of them had dreamed of this— neither of them. What could she say to Lady Agatha? What would Lady Agatha say to her, though it had not been her fault? She had not dreamed that such a thing could be possible. How could she, oh, how could she?

She was standing in the middle of her room with clasped hands. There was a knock upon the door, and Lady Agatha herself came to her.

What had occurred? Something. It was to be seen in the girl's eyes, and in a certain delicate shyness in her manner.

"Something very nice has happened," she said.

"Something nice?" repeated Emily.

Lady Agatha sat down. The letter from Curzon Street was in her hand half unfolded.

"I have had a letter from mamma. It seems almost bad taste to speak of it so soon, but we have talked to each other so much, and you are so kind, that I want to tell you myself. Sir Bruce Norman has been to talk to papa about—about me."

Emily felt that her cup filled to the brim at the moment.

"He is in England again?"

Agatha nodded gently.

"He only went away to—well, to test his own feelings before he spoke. Mamma is delighted with him. I am going home to-morrow."

Emily made a little swoop forward.

"You always liked him?" she said.

Lady Agatha's delicate mounting colour was adorable.

"I was quite *unhappy*," she owned, and hid her lovely face in her hands.

In the morning-room Lord Walderhurst was talking to Lady Maria.

"You need not give Emily Fox-Seton any more clothes, Maria," he said. "I am going to supply her in future. I have asked her to marry me."

Lady Maria lightly gasped, and then began to laugh.

"Well, James," she said, "you have certainly much more sense than most men of your rank and age."

* * * * *

PART TWO

CHAPTER SEVEN

When Miss Emily Fox-Seton was preparing for the extraordinary change in her life which transformed her from a very poor, hardworking woman into one of the richest marchionesses in England, Lord Walderhurst's cousin, Lady Maria Bayne, was extremely good to her. She gave her advice, and though advice is a cheap present as far as the giver is concerned, there are occasions when it may be a very valuable one to the recipient. Lady Maria's was valuable to Emily Fox-Seton, who had but one difficulty, which was to adjust herself to the marvellous fortune which had befallen her.

There was a certain thing Emily found herself continually saying. It used to break from her lips when she was alone in her room, when she was on her way to her dressmaker's, and in spite of herself, sometimes when she was with her whilom patroness.

"I can't believe it is true! I can't believe it!"

"I don't wonder, my dear girl," Lady Maria answered the second time she heard it. "But what circumstances demand of you is that you should learn to."

"Yes," said Emily, "I know I must. But it seems like a dream. Sometimes," passing her hand over her forehead with a little laugh, "I feel as if I should suddenly find myself wakened in the room in Mortimer Street by Jane Cupp bringing in my morning tea. And I can see the wallpaper and the Turkey-red cotton curtains. One of them was an inch or so too short. I never could afford to buy the new bit, though I always intended to."

"How much was the stuff a yard?" Lady Maria inquired.

"Sevenpence."

"How many yards did you need?"

"Two. It would have cost one and twopence, you see. And I really could get on without it."

Lady Maria put up her lorgnette and looked at her protégée with an interest which bordered on affection, it was so enjoyable to her epicurean old mind.

"I didn't suspect it was as bad as that, Emily," she said. "I should never have dreamed it. You managed to do yourself with such astonishing decency. You were actually nice—always."

"I was very much poorer than anyone knew," said Emily. "People don't like one's troubles. And when one is earning one's living as I was, one must be agreeable, you know. It would never do to seem tiresome."

"There's cleverness in realising that fact," said Lady Maria. "You were always the most cheerful creature. That was one of the reasons Walderhurst admired you."

The future marchioness blushed all over. Lady Maria saw even her neck itself blush, and it amused her ladyship greatly. She was intensely edified by the fact that Emily could be made to blush by the mere mention of her mature fiancé's name.

"She's in such a state of mind about the man that she's delightful," was the old woman's internal reflection; "I believe she's in love with him, as if she was a nurse-maid and he was a butcher's boy."

"You see," Emily went on in her nice, confiding way (one of the most surprising privileges of her new position was that it made it possible for her to confide in old Lady Maria), "it was not only the living from day to day that made one anxious, it was the Future!" (Lady Maria knew that the word began in this case with a capital letter.) "No one knows what the Future is to poor women. One knows that one must get older, and one may not keep well, and if one could not be active and in good spirits, if one could not run about on errands, and things fell off, *what* could one do? It takes hard work, Lady Maria, to keep up even the tiniest nice little room and the plainest presentable wardrobe, if one isn't clever. If I had been clever it would have been quite different, I dare say. I have been so frightened sometimes in the middle of the night, when I wakened and thought about living to be sixty-five, that I have lain and shaken all over. You see," her blush had so far disappeared that she looked for the moment pale at the memory, "I had nobody—nobody."

"And now you are going to be the Marchioness of Walderhurst," remarked Lady Maria.

Emily's hands, which rested on her knee, wrung themselves together.

"That is what it seems impossible to believe," she said, "or to be grateful enough for to—to—" and she blushed all over again.

"Say 'James'," put in Lady Maria, with a sinful if amiable sense of comedy; "you will have to get accustomed to thinking of him as 'James' sometimes, at all events."

But Emily did not say "James." There was something interesting in the innocent fineness of her feeling for Lord Walderhurst. In the midst of her bewildered awe and pleasure at the material splendours looming up in her horizon, her soul was filled with a tenderness as exquisite as the religion of a child. It was a combination of intense gratitude and the guileless passion of a hitherto wholly unawakened woman—a woman who had not hoped for love or allowed her thoughts to dwell upon it, and who therefore had no clear understanding of its full meaning. She could not have explained her feeling if she had tried, and she did not dream of trying. If a person less inarticulate than herself had translated it to her she would have been amazed and abashed. So would Lord Walderhurst have been amazed, so would Lady Maria; but her ladyship's amazement would have expressed itself after its first opening of the eyes, with a faint elderly chuckle.

When Miss Fox-Seton had returned to town she had returned with Lady Maria to South Audley Street. The Mortimer Street episode was closed, as was the Cupps' house. Mrs. Cupp and Jane had gone to Chichester, Jane leaving behind her a letter the really meritorious neatness of which was blotted by two or three distinct tears. Jane respectfully expressed her affectionate rapture at the wondrous news which "Modern Society" had revealed to her before Miss Fox-Seton herself had time to do so.

"I am afraid, miss," she ended her epistle, "that I am not experienced enough to serve a lady in a grand position, but hoping it is not a liberty to ask it, if at any time your own maid should be wanting a young woman to work under her, I should be grateful to be remembered. Perhaps having learned your ways, and being a good needlewoman and fond of it, might be a little recommendation for me."

"I *should* like to take Jane for my maid," Emily had said to Lady Maria. "Do you think I might make her do?"

"She would probably be worth half a dozen French minxes who would amuse themselves by getting up intrigues with your footmen," was Lady Maria's astute observation. "I would pay an extra ten pounds a year myself for slavish affection, if it was to be obtained at agency offices. Send her to a French hairdresser to take a course of lessons, and she will be worth anything. To turn you out perfectly will be her life's ambition."

To Jane Cupp's rapture the next post brought her the following letter:—

> DEAR JANE,—It is just like you to write such a nice letter to me, and I can assure you I appreciated all your good wishes very much. I feel that I have been most fortunate, and am, of course, very happy. I have spoken to Lady Maria Bayne about you, and she thinks that you might make me a useful maid if I gave you the advantage of a course of lessons in hairdressing. I myself know that you would be faithful and interested and that I could not have a more trustworthy young woman. If your mother is willing to spare you, I will engage you. The wages would be thirty-five pounds a year (and beer, of course) to begin with, and an increase later as you became more accustomed to your duties. I am glad to hear that your mother is so well and comfortable. Remember me to her kindly.
>
> Yours truly,
Emily Fox-Seton

Jane Cupp trembled and turned pale with joy as she read her letter.

"Oh, mother!" she said, breathless with happiness. "And to think she is almost a marchioness this very minute. I wonder if I shall go with her to Oswyth Castle first, or to Mowbray, or to Hurst?"

"My word!" said Mrs. Cupp, "you are in luck, Jane, being as you'd rather be a lady's maid than live private in Chichester. You needn't go out to service, you know. Your uncle's always ready to provide for you."

"I know he is," answered Jane, a little nervous lest obstacles might be put in the way of her achieving her long-cherished ambition. "And it's kind of him, and I'm sure I'm grateful. But—though I wouldn't hurt his feelings by mentioning it—it is more independent to be earning your own living, and there's more *life*, you see, in waiting on a titled lady and dressing her for drawing-rooms and parties and races and things, and travelling about with her to the grand places she lives in and visits. Why, mother, I've heard tell that the society in the servants' halls is almost like high life. Butlers and footmen and maids to high people has seen so much of the world and get such manners. Do you remember how quiet and elegant Susan Hill was that was maid to Lady Cosbourne? And she'd been to Greece and to India. If Miss Fox-Seton likes travel and his lordship likes it, I may be taken to all sorts of wonderful places. Just think!"

She gave Mrs. Cupp a little clutch in her excitement. She had always lived in the basement kitchen of a house in Mortimer Street and had never had reason to hope she might leave it. And now!

"You're right, Jane!" her mother said, shaking her head. "There's a great deal in it, particular when you're young. There's a great deal in it."

When the engagement of the Marquis of Walderhurst had been announced, to the consternation of many, Lady Maria had been in her element. She was really fine at times in her attitude towards the indiscreetly or tactlessly inquiring. Her management of Lady Malfry in particular had been a delightful thing. On hearing of her niece's engagement, Lady Malfry had naturally awakened to a proper and well-behaved if belated interest in her. She did not fling herself upon her breast after the manner of worldly aunts in ancient comedies in which Cinderella attains fortune. She wrote a letter of congratulation, after which she called at South Audley Street, and with not too great obviousness placed herself and her house at the disposal of such female relatives as required protection during the period of their preparation for becoming marchionesses. She herself could not have explained exactly how it was that, without being put through any particular process, she understood, before her call was half over, that Emily's intention was to remain with

Lady Maria Bayne and that Lady Maria's intention was to keep her. The scene between the three was far too subtle to be of the least use upon the stage, but it was a good scene, nevertheless. Its expression was chiefly, perhaps, a matter of inclusion and exclusion, and may also have been largely telepathic; but after it was over, Lady Maria chuckled several times softly to herself, like an elderly bird of much humour, and Lady Malfry went home feeling exceedingly cross.

She was in so perturbed a humour that she dropped her eyelids and looked rather coldly down the bridge of her nose when her stupidly cheery little elderly husband said to her,—

"Well, Geraldine?"

"I beg pardon," she replied. "I don't quite understand."

"Of course you do. How about Emily Fox-Seton?"

"She seems very well, and of course she is well satisfied. It would not be possible for her to be otherwise. Lady Maria Bayne has taken her up."

"She is Walderhurst's cousin. Well, well! It will be an immense position for the girl."

"Immense," granted Lady Malfry, with a little flush. A certain tone in her voice conveyed that discussion was terminated. Sir George knew that her niece was not coming to them and that the immense position would include themselves but slightly.

Emily was established temporarily at South Audley Street with Jane Cupp as her maid. She was to be married from Lady Maria's lean old arms, so to speak. Her ladyship derived her usual epicurean enjoyment from the whole thing,—from too obviously thwarted mothers and daughters; from Walderhurst, who received congratulations with a civilly inexpressive countenance which usually baffled the observer; from Emily, who was overwhelmed by her emotions, and who was of a candour in action such as might have appealed to any heart not adapted by the flintiness of its nature to the macadamising of roads.

If she had not been of the most unpretentious nice breeding and unaffected taste, Emily might have been ingenuously funny in her process of transformation.

"I keep forgetting that I can afford things," she said to Lady Maria. "Yesterday I walked such a long way to match a piece of silk, and when I was tired I got into a penny bus. I did not remember until it was too late that I ought to have called a hansom. Do you think," a shade anxiously, "that Lord Walderhurst would mind?"

"Just for the present, perhaps, it would be as well that I should see that you shop in the carriage," her ladyship answered with a small grin. "When you are a marchioness you may make penny buses a feature of the distinguished *insouciance* of your character if you like. I shouldn't myself, because they jolt and stop to pick up people, but you can, with originality and distinction, if it amuses you."

"It doesn't," said Emily. "I hate them. I have longed to be able to take hansoms. Oh! how I have *longed*—when I was tired."

The legacy left her by old Mrs. Maytham had been realised and deposited as a solid sum in a bank. Since she need no longer hoard the income of twenty pounds a year, it was safe to draw upon her capital for her present needs. The fact made her feel comfortable. She could make her preparations for the change in her life with a decent independence. She would have been definitely unhappy if she had been obliged to accept favours at this juncture. She felt as if she could scarcely have borne it. It seemed as if everything conspired to make her comfortable as well as blissfully happy in these days.

Lord Walderhurst found an interest in watching her and her methods. He was a man who, in certain respects, knew himself very well and had few illusions respecting his own character. He had always been rather given to matter-of-fact analysis of his own emotions; and at Mallowe he had once or twice asked himself if it was not disagreeably possible that the first moderate glow of his St. Martin's summer might die away and leave him feeling slightly fatigued and embarrassed by the new aspect of his previously regular and entirely self-absorbed existence. You might think that you would like to marry a woman and then you might realise that there were objections—that even the woman herself, with all her desirable qualities, might be an objection in the end, that any woman might be an objection; in fact, that it required an effort to reconcile oneself to the fact of a woman's being continually about. Of course the arriving at such a conclusion, after one had committed oneself, would be annoying. Walderhurst had, in fact, only reflected upon this possible aspect of affairs *before* he had driven over the heath to pick Emily up. Afterwards he had, in some remote portion of his mentality, vaguely awaited developments.

When he saw Emily day by day at South Audley Street, he found he continued to like her. He was not clever enough to analyse her; he could only watch her, and he always looked on at her with curiosity and a novel sensation rather like pleasure. She wakened up at sight of him, when he called, in a way that was attractive even to an unimaginative man. Her eyes seemed to warm, and she often looked flushed and softly appealing. He began to note vaguely that her dresses were better, and oftener changed, than they had been at Mallowe. A more observant man might have been touched by the suggestion that she was unfolding petal by petal like a flower, and that each carefully chosen costume was a new petal. He did not in the least suspect the reverent eagerness of her care of herself as an object hoping to render itself worthy of his qualities and tastes.

His qualities and tastes were of no exalted importance in themselves, but they seemed so to Emily. It is that which by one chance or another so commends itself to a creature as to incite it to the emotion called love, which is really of importance, and which, not speaking in figures, holds the power of life and death. Personality sometimes achieves this, circumstances always aid it; but in all cases the result is the same and sways the world it exists in—during its existence. Emily Fox-Seton had fallen deeply and touchingly in love with this particular prosaic, well-behaved nobleman, and her whole feminine being was absorbed in her adoration of him. Her tender fancy described him by adjectives such as no other human being would have assented to. She felt that he had condescended to her with a generosity which justified worship. This was not true, but it was true for her. As a consequence of this she thought out and purchased her wardrobe with a solemnity of purpose such as might well have been part of a religious ceremonial. When she consulted fashion plates and Lady Maria, or when she ordered a gown at her ladyship's dressmaker's, she had always before her mind, not herself, but the Marchioness of Walderhurst—a Marchioness of Walderhurst whom the Marquis would approve of and be pleased with. She did not expect from him what Sir Bruce Norman gave to Lady Agatha.

Agatha and her lover were of a different world. She saw them occasionally, not often, because the simple selfishness of young love so absorbed them that they could scarcely realise the existence of other persons than themselves. They were to be married, and to depart for fairyland as soon as possible. Both were fond of travel, and when they took ship together their intention was to girdle the world at leisure, if they felt so inclined. They could do anything they chose, and were so blissfully sufficient for each other that there was no reason why they should not follow their every errant fancy.

The lines which had been increasing in Lady Claraway's face had disappeared, and left her blooming with the beauty her daughters had reproduced. This delightful marriage had smoothed away every difficulty. Sir Bruce was the "most charming fellow in England." That fact acted as a charm in itself, it seemed. It was not necessary to go into details as to the mollifying of tradespeople and rearranging of the entire aspect of life at Curzon Street. When Agatha and Emily Fox-Seton met in town for the first time—it was in the drawing room at South Audley Street—they clasped each other's hands with an exchange of entirely new looks.

"You look so—so *well*, Miss Fox-Seton," said Agatha, with actual tenderness.

If she had not been afraid of seeming a little rudely effusive she would have said "handsome" instead of "well," for Emily was sweetly blooming.

"Happiness is becoming to you," she added. "May I say how *glad* I am?"

"Thank you, thank you!" Emily answered. "Everything in the world seems changed, doesn't it?"

"Yes, everything."

They stood and gazed into each other's eyes a few seconds, and then loosed hands with a little laugh and sat down to talk.

It was, in fact, Lady Agatha who talked most, because Emily Fox-Seton led her on and aided her to delicate expansion by her delight in all that in these days made up her existence of pure bliss. It was as if an old-time fairy story were being enacted before Emily's eyes. Agatha without doubt had grown lovelier, she thought; she seemed even fairer, more willowy, the forget-me-not eyes were of a happier blue, as forget-me-nots growing by clear water-sides are bluer than those grown in a mere garden. She appeared, perhaps, even a little taller, and her small head had, if such a thing were possible, a prettier flower-like poise. This, at least, Emily thought, and found her own happiness added to by her belief in her fancy. She felt that nothing was to be wondered at when she heard Agatha speak of Sir Bruce. She could not utter his name or refer to any act of his without a sound in her voice which had its parallel in the light floating haze of blush on her cheeks. In her intercourse with the world in general she would have been able to preserve her customary sweet composure, but Emily Fox-Seton was not the world. She represented a something which was so primitively of the emotions that one's heart spoke and listened to her. Agatha was conscious that Miss Fox-Seton had seen at Mallowe—she could never quite understand how it had seemed so naturally to happen— a phase of her feelings which no one else had seen before. Bruce had seen it since, but only Bruce. There had actually been a sort of confidence between them—a confidence which had been like intimacy, though neither of them had been effusive.

"Mamma is so happy," the girl said. "It is quite wonderful. And Alix and Hilda and Millicent and Eve—oh! it makes such a difference to them. I shall be able," with a blush which expressed a world of relieved affection, "to give them so much pleasure. Any girl who marries happily and—and well—can alter everything for her sisters, if she *remembers*. You see, I shall have reason to remember. I know things from experience. And Bruce is so kind, and gay, and proud of their prettiness. Just imagine their excitement at all being bridesmaids! Bruce says we shall be like a garden of spring flowers. I am so glad," her eyes suddenly quite heavenly in their joyful relief, "that he is *young!*"

The next second the heavenly relieved look died away. The exclamation had been involuntary. It had sprung from her memory of the days when she had dutifully accepted, as her portion, the possibility of being smiled upon by Walderhurst, who was two years older than her father, and her swift realisation of this fact troubled her. It was indelicate to have referred to the mental image even ever so vaguely.

But Emily Fox-Seton was glad too that Sir Bruce was young, that they were all young, and that happiness had come before they had had time to tire of waiting for it. She was so happy herself that she questioned nothing.

"Yes. It is nice," she answered, and glowed with honest sympathy. "You will want to do the same things. It is so agreeable when people who are married like to do the same things. Perhaps you will want to go out a great deal and to travel, and you could not enjoy it if Sir Bruce did not."

She was not reflecting in the least upon domestic circles whose male heads are capable of making themselves extremely nasty under stress of invitations it bores them to accept, and the inclination of wives and daughters to desire acceptance. She was not contemplating with any premonitory regrets a future in which, when Walderhurst did not wish to go out to dinner or disdained a ball, she should stay at home. Far from it. She simply rejoiced with Lady Agatha, who was twenty-two marrying twenty-eight.

"You are not like me," she explained further. "I have had to work so hard and contrive so closely that *everything* will be a pleasure to me. Just to know that I *never* need starve to death or go into the workhouse is such a relief that—"

"Oh!" exclaimed Lady Agatha, quickly and involuntarily laying a hand on hers, startled by the fact that she spoke as if referring to a wholly matter-of-fact possibility.

Emily smiled, realising her feeling.

"Perhaps I ought not to have said that. I forgot. But such things are possible when one is too old to work and has nothing to depend on. You could scarcely understand. When one is very poor one is frightened, because occasionally one cannot help thinking of it."

"But now—now! Oh! how different!" exclaimed Agatha, with heartfelt earnestness.

"Yes. Now I need never be afraid. It makes me so grateful to—Lord Walderhurst."

Her neck grew pink as she said it, just as Lady Maria had seen it grow pink on previous occasions. Moderate as the words were, they expressed ardour.

Lord Walderhurst came in half an hour later and found her standing smiling by the window.

"You look particularly well, Emily. It's that white frock, I suppose. You ought to wear a good deal of white," he said.

"I will," Emily answered. He observed that she wore the nice flush and the soft appealing look, as well as the white frock. "I wish—"

Here she stopped, feeling a little foolish.

"What do you wish?"

"I wish I could do more to please you than wear white—or black—when you like."

He gazed at her, always through the single eyeglass. Even the vaguest approach to emotion or sentiment invariably made him feel stiff and shy. Realising this, he did not quite understand why he rather liked it in the case of Emily Fox-Seton, though he only liked it remotely and felt his own inaptness a shade absurd.

"Wear yellow or pink occasionally," he said with a brief, awkward laugh.

What large, honest eyes the creature had, like a fine retriever's or those of some nice animal one saw in the Zoo!

"I will wear anything you like," she said, the nice eyes meeting his, not the least stupidly, he reflected, though women who were affectionate often looked stupid. "I will do anything you like; you don't know what you have done for me, Lord Walderhurst."

They moved a trifle nearer to each other, this inarticulate pair. He dropped his eyeglass and patted her shoulder.

"Say 'Walderhurst' or 'James'—or—or 'my dear,'" he said. "We are going to be married, you know." And he found himself going to the length of kissing her cheek with some warmth.

"I sometimes wish," she said feelingly, "that it was the fashion to say 'my lord' as Lady Castlewood used to do in 'Esmond.' I always thought it nice."

"Women are not so respectful to their husbands in these days," he answered, with his short laugh. "And men are not so dignified."

"Lord Castlewood was not very dignified, was he?"

He chuckled a little.

"No. But his rank was, in the reign of Queen Anne. These are democratic days. I'll call you 'my lady' if you like."

"Oh! No—no!" with fervour, "I wasn't thinking of anything like that."

"I know you were not," he reassured her. "You are not that kind of woman."

"Oh! how *could* I be?"

"*You* couldn't," good-naturedly. "That's why I like you."

Then he began to tell her his reason for calling at this particular hour. He came to prepare her for a visit from the Osborns, who had actually just returned from India. Captain Osborn had chosen, or chance had chosen for him, this particular time for a long leave. As soon as she heard the name of Osborn, Emily's heart beat a little quickly. She had naturally learned a good deal of detail from Lady Maria since her engagement. Alec Osborn was the man who, since Lord Walderhurst's becoming a widower, had lived in the gradually strengthening belief that the chances were that it would be his enormous luck to inherit the title and estates of the present Marquis of Walderhurst. He was not a very near relation, but he was the next of kin. He was a young man and a strong one, and Walderhurst was fifty-four and could not be called robust. His medical man did not consider him a particularly good life, though he was not often ill.

"He's not the kind of chap who lives to be a hundred and fifty. I'll say that for him," Alec Osborn had said at mess after dinner had made him careless of speech, and he had grinned not too pleasantly when he uttered the words. "The only thing that would completely wipe my eye isn't as likely to happen to him as to most men. He's unsentimental and level headed, and doesn't like marriage. You can imagine how he's chivied by women. A fellow in his position couldn't be let alone. But he doesn't like marriage, and he's a man who knows jolly well what he likes and what he doesn't. The only child died, and if he doesn't marry again, I'm in a safe place. Good Lord! the difference it would make!" and his grin extended itself.

It was three months after this that the Marquis of Walderhurst followed Emily Fox-Seton out upon the heath, and finding her sitting footsore and depressed in spirit beside the basket of Lady Maria's fish, asked her to marry him.

When the news reached him, Alec Osborn went and shut himself up in his quarters and blasphemed until his face was purple and big drops of sweat ran down it. It was black bad luck—it was black bad luck, and it called for black curses. What the articles of furniture in the room in the bungalow heard was rather awful, but Captain Osborn did not feel that it did justice to the occasion.

When her husband strode by her to his apartment, Mrs. Osborn did not attempt to follow him. She had only been married two years, but she knew his face too well; and she also knew too well all the meaning of the fury contained in the words he flung at her as he hurled himself past her.

"Walderhurst is going to be married!" Mrs. Osborn ran into her own room and sat down clutching at her hair as she dropped her face in her little dark hands. She was an Anglo-Indian girl who had never been home, and had not had much luck in life at any time, and her worst luck had been in being handed over by her people to this particular man, chiefly because he was the next of kin to Lord Walderhurst. She was a curious, passionate creature, and had been in love with him in her way. Her family had been poor and barely decently disreputable. She had lived on the outskirts of things, full of intense girlish vanity and yearnings for social recognition, poorly dressed, passed over and snubbed by people she aspired to know socially, seeing other girls with less beauty and temperament enjoying flirtations with smart young officers, biting her tongue out with envy and bitterness of thwarted spirit. So when Captain Osborn cast an eye on her and actually began a sentimental episode, her relief and excitement at finding herself counting as other girls did wrought itself up into a passion. Her people were prompt and sharp enough to manage the rest, and Osborn was married before he knew exactly whither he was tending. He was not pleased with himself when he wakened to face facts. He could only console himself for having been cleverly led and driven into doing the thing he did not want to do, by the facts that the girl was interesting and clever and had a good deal of odd un-English beauty.

It was a beauty so un-English that it would perhaps appear to its greatest advantage in the contrasts afforded by life in England. She was so dark, of heavy hair and drooping-lidded eyes and fine grained skin, and so sinuous of lithe, slim body, that among native beauties she seemed not to be sufficiently separated by marks of race. She had tumbled up from childhood among native servants, who were almost her sole companions, and who had taught her curious things. She knew their stories and songs, and believed in more of their occult beliefs than any but herself knew.

She knew things which made her interesting to Alec Osborn, who had a bullet head and a cruel lower jaw, despite a degree of the ordinary good looks. The fact that his chances were good for becoming Marquis of Walderhurst and taking her home to a life of English luxury and splendour was a thing she never forgot. It haunted her in her sleep. She had often dreamed of Oswyth Castle and of standing amidst great people on the broad lawns her husband had described feelingly during tropical days when they had sat together panting for breath. When there had been mention made of the remote, awful possibility that Walderhurst might surrender to the siege laid to him, she had turned sick at the thought. It made her clench her hands until the nails almost pressed into the skin of her palms. She could not bear it. She had made Osborn burst into a big, harsh laugh one day when she had hinted to him that there were occult things to be done which might prevent ill luck. He had laughed first and scowled afterwards, cynically saying that she might as well be working them up.

He had not come out to India followed by regrets and affection. He had been a black sheep at home, and had rather been hustled away than otherwise. If he had been a more admirable kind of fellow, Walderhurst would certainly have made him an allowance; but his manner of life had been such as the Marquis had no patience with in men of any class, and especially abhorred in men whom the accident of birth connected with good names. He had not been lavish in his demonstrations of interest in the bullet-headed young man. Osborn's personableness was not of a kind attractive to the unbiassed male observer. Men saw his cruel young jowl and low forehead, and noticed that his eyes were small. He had a good, swaggering military figure to which uniform was becoming, and a kind of animal good looks which would deteriorate early. His colour would fix and deepen with the aid of steady daily drinking, and his features would coarsen and blur, until by the time he was forty the young jowl would have grown heavy and would end by being his most prominent feature.

While he had remained in England, Walderhurst had seen him occasionally, and had only remarked and heard unpleasant things of him,—a tendency to selfish bad manners, reckless living, and low flirtation. He once saw him on the top of a bus with his arm round the waist of an awful, giggling shop-girl kind of person, who was adorned with tremendous feathers and a thick fringe coming unfrizzled with the heat and sticking out here and there in straight locks on her moist forehead. Osborn thought that the arm business had been cleverly managed with such furtiveness that no one could see it, but Walderhurst was driving solemnly by in his respectable barouche, and he found himself gazing through his monocle directly at his relative, and seeing, from the street below, the point at which the young man's arm lost itself under the profusely beaded short cape. A dull flush rose to his countenance, and he turned away without showing any sign of recognition; but he was annoyed and disgusted, because this particular kind of blatantly vulgar bad taste was the sort of thing he loathed. It was the sort of thing which made duchesses of women who did alluring "turns" at music halls or sang suggestive songs in comic opera, and transformed into the chatelaines of ancient castles young persons who had presided at the ribbon counter. He saw as little as possible of his heir presumptive after this, and if the truth were told, Captain Alec Osborn was something of a factor in the affair of Miss Emily Fox-Seton. If Walderhurst's infant son had lived, or if Osborn had been a refined, even if dull, fellow, there are ten chances to one his lordship would have chosen no second marchioness.

Captain Osborn's life in India had not ended in his making no further debts. He was not a man to put the brake on in the matter of self-indulgence. He got into debt so long as a shred of credit remained to him, and afterwards he tried to add to his resources by cards and betting at races. He made and lost by turn, and was in a desperate state when he got his leave. He applied for it because he had conceived the idea that his going home as a married man might be a good thing for him. Hester, it seemed not at all improbable, might accomplish something with Walderhurst. If she talked to him in her interesting semi-Oriental way, and was fervid and picturesque in her storytelling, he might be attracted by her. She had her charm, and when she lifted the heavy lids of her long black eyes and fixed her gaze upon her hearer as she talked about the inner side of native life, of which she knew such curious, intimate things, people always listened, even in India, where the thing was not so much of a novelty, and in England she might be a sort of sensation.

Osborn managed to convey to her gradually, by a process of his own, a great deal of what he wanted her to do. During the months before the matter of the leave was quite decided, he dropped a word here and there which carried a good deal of suggestion to a mind used to seizing on passing intimations. The woman who had been Hester's Ayah when she was a child had become her maid. She was a woman with a wide, silent acquaintance with her own people. She was seldom seen talking to anyone and seldom seemed to leave the house, but she always knew everything. Her mistress was aware that if at any time she chose to ask her a question about the secret side of things concerning black or white peoples, she would receive information to be relied upon. She felt that she could have heard from her many things concerning her husband's past, present, and future, and that the matter of the probable succession was fully comprehended by her.

When she called her into the room after recovering outwardly from her hour of desperation, she saw that the woman was already aware of the blow that had fallen upon the household. What they said to each other need not be recorded here, but there was more in the conversation than the mere words uttered, and it was one of several talks held before Mrs. Osborn sailed for England with her husband.

"He may be led into taking into consideration the fact that he has cut the ground from under a fellow's feet and left him dangling in the air," said Osborn to his wife. "Best thing will be to make friends with the woman, hang her!"

"Yes, Alec, yes," Hester Osborn answered, just a little feverishly. "We must make friends with her. They say she is a good sort and was frightfully poor herself."

"She won't be poor now, hang her!" remarked Captain Osborn with added fervour. "I should like to break her neck! I wonder if she rides?"

"I'm sure she has not been well enough off to do anything like that."

"Good idea to begin to teach her." And he laughed as he turned on his heel and began to walk the deck with a fellow passenger.

It was these people Lord Walderhurst had come to prepare her for.

"Maria has told you about them, I know," he said. "I dare say she has been definite enough to explain that I consider Osborn altogether undesirable. Under the veneer of his knowledge of decent customs he is a cad. I am obliged to behave civilly to the man, but I dislike him. If he had been born in a low class of life, he would have been a criminal."

"Oh!" Emily exclaimed.

"Any number of people would be criminals if circumstances did not interfere. It depends a good deal on the shape of one's skull."

"Oh!" exclaimed Emily again, "do you think so?"

She believed that people who were bad were bad from preference, though she did not at all understand the preference. She had accepted from her childhood everything she had ever heard said in a pulpit. That Walderhurst should propound ideas such as ministers of the Church of England might regard as heretical startled her, but he could have said nothing startling enough to shake her affectionate allegiance.

"Yes, I do," he answered. "Osborn's skull is quite the wrong shape."

But when, a short time after, Captain Osborn brought the skull in question into the room, covered in the usual manner with neatly brushed, close-cropped hair, Emily thought it a very nice shape indeed. Perhaps a trifle hard and round-looking and low of forehead, but not shelving or bulging as the heads of murderers in illustrated papers generally did. She owned to herself that she did not see what Lord Walderhurst evidently saw, but then she did not expect of herself an intelligence profound enough to follow his superior mental flights.

Captain Osborn was well groomed and well mannered, and his demeanour towards herself was all that the most conventional could have demanded. When she reflected that she herself represented in a way the possible destruction of his hopes of magnificent fortune, she felt almost tenderly towards him, and thought his easy politeness wonderful. Mrs. Osborn, too! How interesting and how beautiful in an odd way Mrs. Osborn was! Every movement of her exceeding slimness was curiously graceful. Emily remembered having read novels whose heroines were described as "undulating." Mrs. Osborn was undulating. Her long, drooping, and dense black eyes were quite unlike other girls' eyes. Emily had never seen anything like them. And she had such a lonely, slow, shy way of lifting them to look at people. She was obliged to look up at tall Emily. She seemed a schoolgirl as she stood near her. Emily was the kind of mistaken creature whose conscience, awakening to unnecessary remorses, causes its owner at once to assume all the burdens which Fate has laid upon the shoulders of others. She began to feel like a criminal herself, irrespective of the shape of her skull. Her own inordinate happiness and fortune had robbed this unoffending young couple. She wished that it had not been so, and vaguely reproached herself without reasoning the matter out to a conclusion. At all events, she was remorsefully sympathetic in her mental attitude towards Mrs. Osborn, and being sure that she was frightened of her husband's august relative, felt nervous herself because Lord Walderhurst bore himself with undated courtesy and kept his monocle fixed in his eye throughout the interview. If he had let it drop and allowed it to dangle in an unbiassed manner from its cord, Emily would have felt more comfortable, because she was sure his demeanour would have appeared a degree more encouraging to the Osborns.

"Are you glad to be in England again?" she asked Mrs. Osborn.

"I never was here before," answered the young woman. "I have never been anywhere but in India."

In the course of the conversation she explained that she had not been a delicate child, and also conveyed that even if she had been one, her people could not have afforded to send her home. Instinct revealed to Emily that she had not had many of the good things of life, and that she was not a creature of buoyant spirits. The fact that she had spent a good many hours of most of her young days in reflecting on her ill-luck had left its traces on her face, particularly in the depths of her slow-moving, black eyes.

They had come, it appeared, in the course of duty, to pay their respects to the woman who was to be their destruction. To have neglected to do so would have made them seem to assume an indiscreet attitude towards the marriage.

"They can't like it, of course," Lady Maria summed them up afterwards, "but they have made up their minds to lump it as respectably as possible."

"I am *so* sorry for them," said Emily.

"Of course you are. And you will probably show them all sorts of indiscreet kindnesses, but don't be too altruistic, my good Emily. The man is odious, and the girl looks like a native beauty. She rather frightens me."

"I don't think Captain Osborn is odious," Emily answered. "And she *is* pretty, you know. She is frightened of us, really."

Remembering days when she herself had been at a disadvantage with people who were fortunate enough to be of importance, and recalling what her secret tremor before them had been, Emily was very nice indeed to little Mrs. Osborn. She knew from experience things which would be of use to her—things about lodgings and things about shops. Osborn had taken lodgings in Duke Street, and Emily knew the quarter thoroughly. Walderhurst watched her being nice, through his fixed eyeglass, and he decided that she had really a very good manner. Its goodness consisted largely in its directness. While she never brought forth unnecessarily recollections of the days when she had done other people's shopping and had purchased for herself articles at sales marked 11-3/4*d*, she was interestingly free from any embarrassment in connection with the facts. Walderhurst, who had been much bored by himself and other people in time past, actually found that it gave a fillip to existence to look on at a woman who, having been one of the hardest worked of the genteel labouring classes, was adapting herself to the role of marchioness by the simplest of processes, and making a very nice figure at it too, in her entirely unbrilliant way. If she had been an immensely clever woman, there would have been nothing special in it. She was not clever at all, yet Walderhurst had seen her produce effects such as a clever woman might have laboured for and only attained by a stroke of genius. As, for instance, when she had met for the first time after her engagement, a certain particularly detestable woman of rank, to whom her relation to Walderhurst was peculiarly bitter. The Duchess of Merwold had counted the Marquis as her own, considering him fitted by nature to be the spouse of her eldest girl, a fine young woman with projecting teeth, who had hung fire. She felt Emily Fox-Seton's incomprehensible success to be a piece of impudent presumption, and she had no reason to restrain the expression of her sentiments so long as she conveyed them by methods of inference and inclusion.

"You must let me congratulate you very warmly, Miss Fox-Seton," she said, pressing her hand with maternal patronage. "Your life has changed greatly since we last saw each other."

"Very greatly indeed," Emily flushed frankly in innocent gratitude as she answered. "You are very kind. Thank you, thank you."

"Yes, a great change." Walderhurst saw that her smile was feline and asked himself what the woman was going to say next. "The last time we met you called to ask me about the shopping you were to do for me. Do you remember? Stockings and gloves, I think."

Walderhurst observed that she expected Emily to turn red and show herself at a loss before the difficulties of the situation. He was on the point of cutting into the conversation and disposing of the matter himself when he realised that Emily was neither gaining colour nor losing it, but was looking honestly into her Grace's eyes with just a touch of ingenuous regret.

"It was stockings," she said. "There were some marked down to one and elevenpence halfpenny at Barratt's. They were really *quite* good for the price. And you wanted four pairs. And when I got there they were all gone, and those at two and three were not the least bit better. I was so disappointed. It was too bad!"

Walderhurst fixed his monocle firmly to conceal the fact that he was verging upon a cynical grin. The woman was known to be the stingiest of small great persons in London, her economies were noted, and this incident was even better than many others society had already rejoiced over. The picture raised in the minds of the hearers of her Grace foiled in the purchase of stockings marked down to 1*s*. 11-1/2*d*. would be a source of rapture for some time to come. And Emily's face! The regretful kindness of it, the retrospective sympathy and candid feeling! It was incredibly good!

"And she did it quite by accident!" he repeated to himself in his inward glee. "She did it quite by accident! She's not clever enough to have done it on purpose. What a brilliantly witty creature she would be if she had invented it!"

As she had been able unreluctantly to recall her past upon this occasion, so she was able to draw for Mrs. Osborn's benefit from the experience it had afforded her. She wanted to make up to her, in such ways as she could, for the ill turn she had inadvertently done her. As she had at once ranged herself as an aid on the side of Lady Agatha, so she ranged herself entirely without obtrusiveness on the side of the Osborns.

"It's true that she's a good sort," Hester said when they went away. "Her days of being hard up are not far enough away to be forgotten. She hasn't any affectation, at any rate. It makes it easier to stand her."

"She looks like a strong woman," said Osborn. "Walderhurst got a good deal for his money. She'll make a strapping British matron."

Hester winced and a dusky red shot up in her cheek. "So she will," she sighed.

It was quite true, and the truer it was the worse for people who despairingly hung on and were foolish enough to hope against hope.

CHAPTER EIGHT

The marriage of Lady Agatha came first, and was a sort of pageant. The female writers for fashion papers lived upon it for weeks before it occurred and for some time after. There were numberless things to be written about it. Each flower of the garden of girls was to be described, with her bridesmaid's dress, and the exquisite skin and eyes and hair which would stamp her as the beauty of her season when she came out. There yet remained five beauties in Lady Claraway's possession, and the fifth was a baby thing of six, who ravished all beholders as she toddled into church carrying her sister's train, aided by a little boy page in white velvet and point lace.

The wedding was the most radiant of the year. It was indeed a fairy pageant, of youth and beauty, and happiness and hope.

One of the most interesting features of the occasion was the presence of the future Marchioness of Walderhurst, "the beautiful Miss Fox-Seton." The fashion papers were very strenuous on the subject of Emily's beauty. One of them mentioned that the height and pose of her majestic figure and the cut of her profile suggested the Venus of Milo. Jane Cupp cut out every paragraph she could find and, after reading them aloud to her young man, sent them in a large envelope to Chichester. Emily, faithfully endeavouring to adjust herself to the demands of her approaching magnificence, was several times alarmed by descriptions of her charms and accomplishments which she came upon accidentally in the course of her reading of various periodicals.

The Walderhurst wedding was dignified and distinguished, but not radiant. The emotions Emily passed through during the day—from her awakening almost at dawn to the silence of her bedroom at South Audley Street, until evening closed in upon her sitting in the private parlour of an hotel in the company of the Marquis of Walderhurst—it would require too many pages to describe.

Her first realisation of the day brought with it the physical consciousness that her heart was thumping—steadily thumping, which is quite a different matter from the ordinary beating—at the realisation of what had come at last. An event which a year ago the wildest dream could not have depicted for her was to-day an actual fact; a fortune such as she would have thought of with awe if it had befallen another woman, had befallen her unpretending self. She passed her hand over her forehead and gasped as she thought of it.

"I hope I shall be able to get accustomed to it and not be a—a disappointment," she said. "Oh!" with a great rising wave of a blush, "how good of him! How can I ever—"

She lived through the events of the day in a sort of dream within a dream. When Jane Cupp brought her tea, she found herself involuntarily making a mental effort to try to look as if she was really awake. Jane, who was an emotional creature, was inwardly so shaken by her feelings that she herself had stood outside the door a few moments biting her lips to keep them from trembling, before she dared entirely trust herself to come in. Her hand was far from steady as she set down the tray.

"Good morning, Jane," Emily said, by way of trying the sound of her voice.

"Good morning, miss," Jane answered. "It's a beautiful morning, miss. I hope—you are very well?"

And then the day had begun.

Afterwards it marched on with solemn thrill and stately movement through hours of wondrous preparation for an imposing function, through the splendid gravity of the function itself, accompanied by brilliant crowds collected and looking on in a fashionable church, and motley crowds collected to look on outside the edifice, the latter pushing and jostling each other and commenting in more or less respectful if excited undertones, but throughout devouring with awe-struck or envious eyes. Great people whom Emily had only known through the frequent mention of their names in newspapers or through their relationship or intimacy with her patrons, came to congratulate her in her rôle of bride. She seemed to be for hours the centre of a surging, changing crowd, and her one thought was to bear herself with an outward semblance of composure. No one but herself could know that she was saying internally over and over again, to steady herself, making it all seem real, "I am being married. This is my wedding. I am Emily Fox-Seton being married to the Marquis of Walderhurst. For his sake I must not look stupid or excited. I am not in a dream."

How often she said this after the ceremony was over and they returned to South Audley Street, for the wedding breakfast could scarcely be computed. When Lord Walderhurst helped her from the carriage and she stepped on to the strip of red carpet and saw the crowd on each side of it and the coachman and footmen with their big white wedding favours and the line of other equipages coming up, her head whirled.

"That's the Marchioness," a young woman with a bandbox exclaimed, nudging her companion. "That's 'er! Looks a bit pale, doesn't she?"

"But, oh Gawd! look at them di-monds an' pearls—jess look at 'em!" cried the other. "Wish it was me."

The breakfast seemed splendid and glittering and long; people seemed splendid and glittering and far off; and by the time Emily went to change her bridal magnificence for her travelling costume she had borne as much strain as she was equal to. She was devoutly grateful for the relief of finding herself alone in her bedroom with Jane Cupp.

"Jane," she said, "you know exactly how many minutes I can dress in and just when I must get into the carriage. Can you give me five minutes to lie down quite flat and dab my forehead with eau de cologne? Five minutes, Jane. But be quite sure."

"Yes, miss—I do beg pardon—my lady. You can have five—safe."

She took no more,—Jane went into the dressing-room and stood near its door, holding the watch in her hand,—but even five minutes did her good.

She felt less delirious when she descended the stairs and passed through the crowds again on Lord Walderhurst's arm. She seemed to walk through a garden in resplendent bloom. Then there were the red carpet once more, and the street people, and the crowd of carriages and liveries, and big, white favours.

Inside the carriage, and moving away to the echo of the street people's cheer, she tried to turn and look at Lord Walderhurst with an unalarmed, if faint, smile.

"Well," he said, with the originality which marked him, "it is really over!"

"Yes," Emily agreed with him. "And I never can forget Lady Maria's goodness."

Walderhurst gazed at her with a dawning inquiry in his mind. He himself did not know what the inquiry was. But it was something a trifle stimulating. It had something to do with the way in which she had carried herself throughout the whole thing. Really few women could have done it as well. The pale violet of her travelling costume which was touched with sable was becoming to her fine, straight figure. And at the moment her eyes rested on his with the suggestion of trustful appeal. Despite the inelasticity of his mind, he vaguely realised his bridegroom honours.

"I can begin now," he said with stiff lightness, if such a paradox can be, "to address you as the man in Esmond addressed his wife. I can call you 'my lady.'"

"Oh!" she said, still trying to smile, but quivering.

"You look very nice," he said. "Upon my word you do."

And kissed her trembling honest mouth almost as if he had been a man—not quite—but almost.

CHAPTER NINE

They began the new life at Palstrey Manor, which was ancient and most beautiful. Nothing Walderhurst owned was as perfect an example of olden time beauty, and as wonderful for that reason. Emily almost wept before the loveliness of it, though it would not have been possible for her to explain or particularise the grounds for her emotion. She knew nothing whatever of the venerable wonders of the architecture. To her the place looked like an immense, low-built, rambling fairy palace—the palace of some sleeping beauty during whose hundred years of slumber rich dark-green creepers had climbed and overgrown its walls and towers, enfolding and festooning them with leaves and tendrils and actual branches. The huge park held an enchanted forest of trees; the long avenue of giant limes, their writhen limbs arching and interlocking, their writhen roots deep in velvet moss, was an approach suited to a fairy story.

* * * * *

During her first month at Palstrey Emily went about still in her dream. It became more a dream every day. The old house was part of it, the endless rooms, the wonderful corridors, the gardens with their revelations of winding walks, labyrinths of evergreens, and grass paths leading into beautiful unexpected places, where one suddenly came upon deep, clear pools where water plants grew and slow carp had dreamed centuries away. The gardens caused Emily to disbelieve in the existence of Mortimer Street, but the house at times caused her to disbelieve in herself. The picture gallery especially had this effect upon her. The men and women, once as alive as her everyday self, now gazing down at her from their picture frames sometimes made her heart beat as if she stood in the presence of things eerie. Their strange, rich, ugly, or beautiful garments, their stolid or fervid, ugly or beautiful, faces, seemed to demand something of her; at least she had just enough imagination to feel somewhat as if they did. Walderhurst was very kind to her, but she was afraid she might bore him by the exceeding ignorance of her questions about people whom he had known from his childhood as his own kith and kin. It was not unlikely that one might have become so familiar with a man in armour or a woman in a farthingale that questions connected with them might seem silly. Persons whose ancestors had always gazed intimately at them from walls might not unnaturally forget that there were other people to whom they might wear only the far-away aspect of numbers in catalogues of the Academy, or exhibitions of that order.

There was a very interesting catalogue of the Palstrey pictures, and Emily found and studied it with deep interest. She cherished a touching secret desire to know what might be discoverable concerning the women who had been Marchionesses of Walderhurst before. None of them but herself, she gathered, had come to their husbands from bed-sitting rooms in obscure streets. There had been noble Hyrsts in the reign of Henry I., and the period since then elapsed had afforded time for numerous bridals. Lady Walderhurst was overcome at moments by her reflections upon what lay behind and before her, but not being a complex person or of fervid imagination, she was spared by nature the fevers of complex emotions.

In fact, after a few weeks had passed she came out of her dream and found her happiness enduring and endurable. Each day's awakening was a delight to her, and would probably be so to the end of her existence, absolutely because she was so sane and uncomplex a creature. To be deftly assisted in her dressing by Jane Cupp, and to know that each morning she might be fittingly and becomingly attired without anxiety as to where her next gown was to come from, was a lovely thing. To enjoy the silent, perfect workings of the great household, to drive herself or be driven, to walk and read, to loiter through walled gardens and hothouses at will,—such things to a healthy woman with an unobscured power of enjoyment were luxuries which could not pall.

Walderhust found her an actual addition to his comfort. She was never in the way. She seemed to have discovered the trick of coming and going undisturbingly. She was docile and affectionate, but not in the least sentimental. He had known men whose first years of marriage, not to speak of the first months, had been rendered unbearable by the fact that their wives were constantly demanding or expecting the expression of sentiments which unsentimental males had not at their fingers' ends. So the men had been annoyed or bored, and the women had been dissatisfied. Emily demanded nothing of the sort, and was certainly not dissatisfied. She looked very handsome and happy. Her looks positively improved, and when people began to call and she to pay visits, she was very much liked. He had certainly been quite right in deciding to ask her to marry him. If she had a son, he should congratulate himself greatly. The more he saw of Osborn the more he disliked him. It appeared that there was a prospect of a child there.

This last was indeed true, and Emily had been much touched and awakened to sympathy. It had gradually become revealed to her that the Osborns were poorer than they could decently admit. Emily had discovered that they could not even remain in the lodgings in Duke Street, though she did not know the reason, which was that Captain Osborn had been obliged to pay certain moneys to stave off a scandal not entirely unconnected with the young woman his arm had encircled the day Walderhurst had seen him on the top of the bus. He was very well aware that if he was to obtain anything from Lord Walderhurst, there were several things which must be kept entirely dark. Even a scandal belonging to the past could be made as unpleasant as an error of to-day. Also the young woman of the bead cape knew how to manage him. But they must remove to cheaper lodgings, and the rooms in Duke Street had been far from desirable.

Lady Walderhurst came in one morning from a walk, with a fresh colour and bright eyes, and before taking off her hat went to her husband's study.

"May I come in?"

Walderhurst had been writing some uninteresting letters and looked up with a smile.

"Certainly," he answered. "What a colour you have! Exercise agrees with you. You ought to ride."

"That was what Captain Osborn said. If you don't mind, I should like to ask you something."

"I don't mind. You are a reasonable woman, Emily. One's safe with you."

"It is something connected with the Osborns."

"Indeed!" chilling slightly. "I don't care about them, you know."

"You don't dislike her, do you?"

"No-o, not exactly."

"She's—the truth is, she is not at all well," with a trifle of hesitance; "she ought to be better taken care of than she is in lodgings, and they are obliged to take very cheap ones."

"If he had been a more respectable fellow his circumstances would have been different," rather stiffly.

Emily felt alarmed. She had not dreamed of the temerity of any remark suggestive of criticism.

"Yes," hastily, "of course. I am sure you know best; but—I thought perhaps—"

Walderhurst liked her timidity. To see a fine, tall, upstanding creature colour in that way was not disagreeable when one realised that she coloured because she feared she might offend one.

"What did you think 'perhaps'?" was his lenient response.

Her colour grew warmer, but this time from a sense of relief, because he was evidently not as displeased as he might have been.

"I took a long walk this morning," she said. "I went through the High Wood and came out by the place called The Kennel Farm. I was thinking a good deal of poor Mrs. Osborn because I had heard from her this morning, and she seemed so unhappy. I was looking at her letter again when I turned into the lane leading to the house. Then I saw that no one was living there, and I could not help going in to look—it is such a delightful old building, with its queer windows and chimneys, and the ivy which seems never to have been clipped. The house is so roomy and comfortable—I peeped in at windows and saw big fireplaces with benches inside them. It seems a pity that such a place should not be lived in and—well, I thought how *kind* it would be of you to lend it to the Osborns while they are in England."

"It would indeed be kind," remarked his lordship, without fervour.

Her momentary excitement led Emily to take the liberty of putting out her hand to touch his. She always felt as if connubial familiarities were rather a liberty; at least she had not, so far, been able to overcome a feeling rather of that order. And this was another thing Walderhurst by no means disliked. He himself was not aware that he was a man with a good deal of internal vanity which enjoyed soothing food. In fact, he had not a sufficiently large brain to know very much about himself or to be able to analyse his reasons for liking or disliking people or things. He thought he knew his reasons for his likes and dislikes, but he was frequently very far away from the clear, impersonal truth about them. Only the brilliant logic and sensitiveness of genius really approaches knowledge of itself, and as a result it is usually extremely unhappy. Walderhurst was never unhappy. He was sometimes dissatisfied or annoyed, but that was as far as his emotions went.

Being pleased by the warm touch of Emily's hand, he patted her wrist and looked agreeably marital.

"The place was built originally for a family huntsman, and the pack was kept there. That is why it is called The Kennel Farm. When the last lease fell out it remained unlet because I don't care for an ordinary tenant. It's the kind of house that is becoming rare, and the bumpkin farmer and his family don't value antiquities."

"If it were furnished as it *could* be furnished," said Emily, "it would be *beautiful*. One *can* get old things in London if one can afford them. I've seen them when I've been shopping. They are not cheap, but you can get them if you really search."

"Would you like to furnish it?" Walderhurst inquired. The consciousness that he could, if he chose, do the utmost thing of its kind in this way, at the moment assumed a certain proportion of interest to him under the stimulation of the wonder and delight which leaped into Emily's eyes as the possibility confronted her. Having been born without imagination, his wealth had not done for him anything out of the ordinary every-day order.

"Would I *like* to do it? Oh, *dear!*" she exclaimed. "Why, in all my life I have never *dreamed* of being able to do such things."

That, of course, was true, he reflected, and the fact added to his appreciation of the moment. There were, of course, many people to whom it would be impossible to contemplate the spending of a sum of money of any importance in the indulgence of a wish founded on mere taste. He had not thought of the thing particularly in detail before, and now that he realised the significance of the fact as a fact, Emily had afforded him a new sensation.

"You may do it now, if you wish," he said. "I once went over the place with an architect, and he said the whole thing could be made comfortable and the atmosphere of the period wholly retained for about a thousand pounds. It is not really dilapidated and it is worth saving. The gables and chimneys are very fine. I will attend to that, and you can do the rest in your own way."

"It may take a good deal of money to buy the old things," gasped Emily. "They are not cheap in these days. People have found out that they are wanted."

"It won't cost twenty thousand pounds," Walderhurst answered. "It is a farm-house after all, and you are a practical woman. Restore it. You have my permission."

Emily put her hands over her eyes. This was being the Marchioness of Walderhurst, and made Mortimer Street a thing still more incredible. When she dropped her hands, she laughed even a trifle hysterically.

"I *couldn't* thank you," she said. "It is as I said. I never quite believed there were people who were able to think of doing such things."

"There are such people," he said. "You are one of them."

"And—and—" She put it to him with a sudden recollection of the thing her emotions had momentarily swept away. "Oh! I must not forget, because I am so pleased. When it is furnished—"

"Oh! the Osborns? Well, we will let them have it for a few months, at any rate."

"They will be so *thankful*," emotionally. "You will be doing them *such* a favour."

"I am doing it for you, not for them. I like to see you pleased."

She went to take off her hat with moisture in her eyes, being overpowered by his munificence. When she reached her room she walked about a little, because she was excited, and then sat down to think of the relief her next letter would carry to Mrs. Osborn. Suddenly she got up, and, going to her bedside, knelt down. She respectfully poured forth devout thanks to the Deity she appealed to when she aided in the intoning of the Litany on Sundays. Her conception of this Power was of the simplest conventional nature.

She would have been astonished and frightened if she had been told that she regarded the Omnipotent Being as possessing many of the attributes of the Marquis of Walderhurst. This was, in fact, true without detracting from her reverence in either case.

CHAPTER TEN

The Osborns were breakfasting in their unpleasant sitting-room in Duke Street when Lady Walderhurst's letter arrived. The toast was tough and smoked, and the eggs were of the variety labelled "18 a shilling" in the shops; the apartment was also redolent of kippered herring, and Captain Osborn was scowling over the landlady's weekly bill when Hester opened the envelope stamped with a coronet. (Each time Emily wrote a note and found herself confronting the coronet on the paper, she blushed a little and felt that she must presently awake from her dream.) Mrs. Osborn herself was looking far from amiable. She was ill and nervous and irritable, and had, in fact, just been crying and wishing that she was dead, which had given rise to unpleasantness between herself and her husband, who was not in the mood to feel patient with nerves.

"Here's one from the Marchioness," she remarked slightingly.

"I have had none from the Marquis," sneered Osborn. "He might have condescended a reply—the cold-blooded beggar!"

Hester was reading her letter. As she turned the first page her expression changed. As has previously been suggested, the epistolary methods of Lady Walderhurst were neither brilliant nor literary, and yet Mrs. Osborn seemed to be pleased by what she read. During the reading of a line or so she wore an expression of slowly questioning wonder, which, a little later on, settled into relief.

"I can only say I think it's very decent of them," she ejaculated at last; "really decent!"

Alec Osborn looked up, still scowlingly.

"I don't see any cheque," he observed. "That would be the most decent thing. It's the thing we want most, with this damned woman sending in bills like this for the fourth-rate things we live on, and for her confounded tenth-rate rooms."

"This is better than cheques. It means our having something we couldn't hope for cheques enough to pay for. They are offering to lend us a beautiful old place to live in for the rest of our stay."

"What!" Osborn exclaimed. "Where?"

"Near Palstrey Manor, where they are staying now."

"Near Palstrey! How near?" He had been slouching in his chair and now sat up and leaned forward on the table. He was eager.

Hester referred to the letter again.

"She doesn't say. It is a sort of antiquity, I gather. It's called The Kennel Farm. Have you ever been to Palstrey?"

"Not as a guest." He was generally somewhat sardonic when he spoke of anything connected with Walderhurst. "But once I was in the nearest county town by chance and rode over. By Jove!" starting a little, "I wonder if it can be a rum old place I passed and reined in to have a look at. I hope it is."

"Why?"

"It's near enough to the Manor to be convenient."

"Do you think," hesitating, "that we shall see much of them?"

"We shall if we manage things decently. She likes you, and she's the kind of woman to be sympathising and make a fuss over another woman—particularly one who is under the weather and can be sentimentalised over."

Hester was pushing crumbs about on the tablecloth with her knife, and a dull red showed itself on her cheek.

"I am not going to make capital of—circumstances," she said sullenly. "I won't."

She was not a woman easily managed, and Osborn had had reason on more than one occasion to realise a certain wicked stubbornness in her. There was a look in her eye now which frightened him. It was desperately necessary that she should be kept in a tractable mood. As she was a girl with affections, and he was a man without any, he knew what to do.

He got up and went to her side, putting his arm round her shoulders as he sat in a chair near her. "Now, little woman," he said. "Now! For God's sake don't take it that way. Don't think I don't understand how you feel."

"I don't believe you know anything about the way I feel," she said, setting her narrow white teeth and looking more like a native woman than he had ever seen her. A thing which did not aid his affection for her, such as it was, happened to be that in certain moods she suggested a Hindoo beauty to him in a way which brought back to him memories of the past he did not care to have awakened.

"Yes I do, yes I do," he protested, getting hold of her hand and trying to make her look at him. "There are things such a woman as you can't help feeling. It's because you feel them that you must be on your mettle—Lord knows you've got pluck enough—and stand by a fellow now. What shall I do, my God, if you don't?"

He was, in fact, in such straits that the ring of emotion in his voice was not by any means assumed.

"My God!" he repeated, "what shall we all do if you won't?"

She lifted her eyes then to look at him. She was in a sufficiently nervous condition to be conscious that tears were always near.

"Are there worse things than you have told me?" she faltered.

"Yes, worse things than it would be fair to bother you with. I don't want you to be tormented. I was a deuced fool before I met you and began to run straight. Things pile in now that would have lain quiet enough if Walderhurst had not married. Hang it all! he ought to do the decent thing by me. He owes something to the man who may stand in his shoes, after all."

Hester lifted her slow eyes again.

"You've not much of a chance now," she said. "She's a fine healthy woman."

Osborn sprang up and paced the floor, set upon by a sudden spasm of impotent rage. He snapped his teeth rather like a dog.

"Oh! curse her!" he gave forth. "The great, fresh-coloured lumping brute! What did she come into it for? Of all the devilish things that can happen to a man, the worst is to be born to the thing I was born to. To know through your whole life that you're just a stone's-throw from rank and wealth and splendour, and to have to live and look on as an outsider. Upon my word, I've felt more of an outsider just because of it. There's a dream I've had every month or so for years. It's a dream of opening a letter that tells me he's dead, or of a man coming into the room or meeting me in the street and saying suddenly, 'Walderhurst died last night, Walderhurst died last night!' They're always the same words, 'Walderhurst died last night!'

And I wake up shaking and in a cold sweat for joy at the gorgeous luck that's come at last."

Hester gave a low cry like a little howl, and dropped her head on her arms on the table among the cups and saucers.

"She'll have a son! She'll have a son!" she cried. "And then it won't matter whether *he* dies or not."

"Ough!" was the sound wrenched from Osborn's fury. "And our son might have been in it. Ours might have had it all! Damn—damn!"

"He won't,—he won't now, even if he lives to be born," she sobbed, and clutched at the dingy tablecloth with her lean little hands.

It was hard on her. She had had a thousand feverish dreams he had never heard of. She had lain awake hours at night and stared with wide-open eyes at the darkness, picturing to her inner soul the dream of splendour that she would be part of, the solace for past miseries, the high revenges for past slights that would be hers after the hour in which she heard the words Osborn had just quoted, "Walderhurst died last night!" Oh! if luck had only helped them! if the spells her Ayah had taught her in secret had only worked as they would have worked if she had been a native woman and had really used them properly! There was a spell she had wrought once which Ameerah had sworn to her was to be relied on. It took ten weeks to accomplish its end. In secret she had known of a man on whom it had been worked. She had found out about it partly from the remote hints which had aided her half knowledge of strange things and by keeping a close watch. The man had died—he had died. She herself, and with her own eyes had seen him begin to ail, had heard of his fevers and pains and final death. He had died. She knew that. And she had tried the thing herself in dead secrecy. And at the fifth week, just as with the native who had died, she heard that Walderhurst was ill. During the next four weeks she was sick with the tension of combined horror and delight. But he did not die in the tenth week. They heard that he had gone to Tangiers with a party of notable people, and that his "slight" indisposition had passed, leaving him in admirable health and spirits.

Her husband had known nothing of her frenzy. She would not have dared to tell him. There were many things she did not tell him. He used to laugh at her native stories of occult powers, though she knew that he had seen some strange things done, as most foreigners had. He always explained such things contemptuously on grounds which presupposed in the performers of the mysteries powers of agility, dexterity, and universal knowledge quite as marvellous as anything occult could have been. He did not like her to show belief in the "tricks of the natives," as he called them. It made a woman look a fool, he said, to be so credulous.

During the last few months a new fever had tormented her. Feelings had awakened in her which were new. She thought things she had never thought before. She had never cared for children or suspected herself of being the maternal woman. But Nature worked in her after her weird fashion. She began to care less for some things and more for others. She cared less for Osborn's moods and was better able to defy them. He began to be afraid of her temper, and she began to like at times to defy his. There had been some fierce scenes between them in which he had found her meet with a flare of fury words she would once have been cowed by. He had spoken one day with the coarse slightingness of a selfish, irritable brute, of the domestic event which was before them. He did not speak twice.

She sprang up before him and shook her clenched fist in his face, so near that he started back.

"Don't say a word!" she cried. "Don't dare—don't dare. I tell you—look out, if you don't want to be killed."

During the outpouring of her frenzy he saw her in an entirely new light and made discoveries. She would fight for her young, as a tigress fights for hers. She was nursing a passion of secret feeling of which he had known nothing. He had not for a moment suspected her of it. She had not seemed that kind of girl. She had been of the kind that cares for finery and social importance and the world's favour, not for sentiments.

On this morning of the letter's arrival he watched her sobbing and clutching the tablecloth, and reflected. He walked up and down and pondered. There were a lot of things to be thought over.

"We may as well accept the invitation at once," he said. "Grovel as much as you choose. The more the better. They'll like it."

CHAPTER ELEVEN

The Osborns arrived at The Kennel Farm on a lovely rainy morning. The green of the fields and trees and hedges was sweetly drenched, and the flowers held drops which sparkled when the fitful sun broke forth and searched for the hidden light in them. A Palstrey carriage comfortably met them and took them to their destination.

As they turned into the lane, Osborn looked out at the red gables and chimneys showing themselves among the trees.

"It's the old place I looked at," he said, "and a jolly old place it is."

Hester was drinking in the pure sweetness of the fresh air and filling her soul with the beauty of such things as she had never seen before. In London she had grown hopeless and sick of spirit. The lodgings in Duke Street, the perpetual morning haddock and questionable eggs and unpaid bills, had been evil things for her. She had reached a point at which she had felt she could bear them no longer. Here, at all events, there would be green trees and clear air, and no landlady. With no rent to pay, there would be freedom from one torment at least.

She had not expected much more than this freedom, however. It had seemed highly probable that there might be discomforts in an ancient farmhouse of the kind likely to be lent to impecunious relatives.

But before they crossed the threshold it was plain to her that, for some reason, they had been given more. The old garden had been put in order—a picturesque and sweet disorderly order, which had allowed creepers to luxuriate and toss, and flowers to spring out of crannies, and clumps of things to mass themselves without restraint.

The girl's wretched heart lifted itself as they drove up to the venerable brick porch which had somewhat the air of a little church vestibule. Through the opened door she saw a quaint comfort she had not dreamed of. She had not the knowledge of things which would have told her what wonders Emily had done with the place, but she could see that its quaint furnishings were oddly beautiful in their harmony. The heavy chairs and benches and settles seemed to have been part of centuries of farm-house life, and to belong to the place as much as the massive beams and doors.

Hester stood in the middle of the hall and looked about her. Part of it was oak panelled and part was whitewashed. There were deep, low windows cut in the thick walls.

"I never saw anything the least like it," she said.

"You wouldn't expect to see anything like it in India," her husband answered. "And you won't find many places like it in England. I should like a look at the stables."

He went out almost immediately and took the look in question, finding the result unexpectedly satisfactory. Walderhurst had lent him a decent horse to ride, and there was a respectable little cart for Hester. Palstrey Manor had "done them" very well. This was a good deal more than he had expected. He knew such hospitality would not have been shown him if he had come to England unmarried. Consequently his good luck was partly a result of Hester's existence in his life. At the same time there awakened in him a consciousness that Hester would not have been likely to produce such results unless in combination with another element in the situation,—the element of another woman who was sympathetic and had some power,—the new Lady Walderhurst, in fact.

"And yet, confound her—confound her!" he thought, as he walked into the loose box to look the mare over and pat her sleekness.

The relations which established themselves between Palstrey and The Kennel Farm were marked by two characteristic features. One of these was that Lord Walderhurst did not develop any warmer interest in the Osborns, and that Lady Walderhurst did. Having acceded to Emily's wishes, and really behaved generously in the matter of providing for his heir presumptive and his wife, Lord Walderhurst felt impelled to no further demonstration of feeling.

"I don't like him any better than I did," he remarked to Emily. "And I cannot say that Mrs. Osborn attracts me. Of course there is a reason why a kind-hearted woman like yourself should be specially good to her just now. Do anything you wish for them while they are in the neighbourhood. But as for me, the fact that a man is one's heir presumptive is not enough in itself alone to endear him to one, rather the contrary."

Between these two it is to be confessed there existed that rancour which is not weakened by the fact that it remains unexpressed and lurks in the deeps of the inward being. Walderhurst would not have been capable of explaining to himself that the thing he chiefly disliked in this robust, warm-blooded young man was that when he met him striding about with his gun over his shoulder and a keeper behind him, the almost unconscious realisation of the unpleasant truth that he was striding over what might prove to be his own acres, and shooting birds which in the future he would himself possess the right to preserve, to invite other people to shoot, to keep less favoured persons from shooting, as lord of the Manor. This was a truth sufficiently irritating to accentuate all his faults of character and breeding.

Emily, whose understanding of his nature developed with every day of her life, grew into a comprehension of this by degrees. Perhaps her greatest leap forward was taken on the day when, as he was driving her in the cart which had picked her up on the moor, they saw Osborn tramping through a cover with his gun. He did not see them, and a shade of irritation swept Walderhurst's face.

"He seems to feel very much at home," he commented.

Then he was silent for a space during which he did not look pleased.

"If he were my son," he said, "it would be a different matter. If Audrey's child had lived—"

He stopped and gave the tall mare a light cut with his whip. He was evidently annoyed with himself for having spoken.

A hot wave of colour submerged Emily. She felt it rush over her whole body. She turned her face away, hoping Walderhurst would not observe her. This was the first time she had heard him utter his dead wife's name. She had never heard anyone speak it. Audrey had evidently not been a much-beloved or regretted person. But she had had a son.

Her primitive soul had scarcely dared to approach, even with awe, the thought of such a possibility for herself. As in the past she had not had the temerity to dream of herself as a woman who possessed attractions likely to lead to marriage, so she was mentally restrained in these days. There was something spinster-like in the tenor of her thoughts. But she would have laid down her life for this dull man's happiness. And of late she had more than once blamed herself for accepting so much, unthinkingly.

"I did not realise things properly," she had said to herself in humble pain. "I ought to have been a girl, young and strong and beautiful. His sacrifice was too great, it was immense."

It had been nothing of the sort. He had pleased himself and done what was likely to tend, and had tended, altogether to his own ease and comfort. In any case Emily Fox-Seton was a fine creature, and only thirty-four, and with Alec Osborn at the other side of the globe the question of leaving an heir had been less present and consequently had dwindled in importance.

The nearness of the Osborns fretted him just now. If their child was a son, he would be more fretted still. He was rather glad of a possibility, just looming, of his being called away from England through affairs of importance.

He had spoken to Emily of this possibility, and she had understood that, as his movements and the length of his stay would be uncertain, she would not accompany him.

"There is one drawback to our marriage," he said.

"Is it—is it anything I can remove?" Emily asked.

"No, though you are responsible for it. People seldom can remove the drawbacks they are responsible for. You have taught me to miss you.",

"Have I—have I?" cried Emily. "Oh! I *am* happy!"

She was so happy that she felt that she must pass on some of her good fortune to those who had less. She was beautifully kind to Hester Osborn. Few days passed without the stopping of a Walderhurst carriage before the door of The Kennel Farm. Sometimes Emily came herself to take Mrs. Osborn to drive, sometimes she sent for her to come to lunch and spend the day or night at Palstrey. She felt an interest in the young woman which became an affection. She would have felt interested in her if there had not existed a special reason to call forth sympathy. Hester had many curious and new subjects for conversation. Emily liked her descriptions of Indian life and her weird little stories of the natives. She was charmed with Ameerah, whose nose rings and native dress, combining themselves with her dark mystic face, rare speech, and gliding, silent movements, awakened awe in the rustics and mingled distrust and respect in the servants' hall at Palstrey.

"She's most respectably behaved, my lady, though foreign and strange in her manners," was Jane Cupp's comment. "But she has a way of looking at a person—almost stealthy—that's upset me many a time when I've noticed it suddenly. They say that she knows things, like fortune-telling and spells and love potions. But she will only speak of them quite secret."

Emily gathered that Jane Cupp was afraid of the woman, and kept a cautious eye upon her.

"She is a very faithful servant, Jane," she answered. "She is devoted to Mrs. Osborn."

"I am sure she is, my lady. I've read in books about the faithfulness of black people. They say they're more faithful than white ones."

"Not more faithful than *some* white ones," said Lady Walderhurst with her good smile. "Ameerah is not more faithful than you, I'm very sure."

"Oh, my lady!" ejaculated Jane, turning red with pleasure. "I do hope not. I shouldn't like to think she could be."

In fact the tropic suggestion of the Ayah's personality had warmed the imagination of the servants' hall, and there had been much talk of many things, of the Osborns as well as of their servants, and thrilling stories of East Indian life had been related by Walderhurst's man, who was a travelled person. Captain Osborn had good sport on these days, and sport was the thing he best loved. He was of the breed of man who can fish, hunt, or shoot all day, eat robust meals and sleep heavily all night; who can do this every day of a year, and in so doing reach his highest point of desire in existence. He knew no other aspirations in life than such as the fortunes of a man like Walderhurst could put him in possession of. Nature herself had built him after the model of the primeval type of English country land-owner. India with her blasting and stifling hot seasons and her steaming rains gave him nothing that he desired, and filled him with revolt against Fate every hour of his life. His sanguine body loathed and grew restive under heat. At The Kennel Farm, when he sprang out of his bed in the fresh sweetness of the morning and plunged into his tub, he drew every breath with a physical rapture. The air which swept in through the diamond-paned, ivy-hung casements was a joy.

"Good Lord!" he would cry out to Hester through her half-opened door, "what mornings! how a man *lives* and feels the blood rushing through his veins! Rain or shine, it's all the same to me. I can't stay indoors. Just to tramp through wet or dry heather, or under dripping or shining trees, is enough. How can one believe one has ever lain sweating with one's tongue lolling out, and listened to the whining creak of the punkah through nights too deadly hot to sleep in! It's like remembering hell while one lives in Paradise."

"We shan't live in Paradise long," Hester said once with some bitterness. "Hell is waiting for us."

"Damn it! don't remind a man. There are times when I don't believe it." He almost snarled the answer. It was true that his habit was to enhance the pleasure of his days by thrusting into the background all recollections of the reality of any other existence than that of the hour. As he tramped through fern and heather he would remember nothing but that there was a chance—there was chance, good Lord! After a man not over strong reached fifty-four or five, there were more chances than there had been earlier.

After hours spent in such moods, it was not pleasant to come by accident upon Walderhurst riding his fine chestnut, erect and staid, and be saluted by the grave raising of his whip to his hat. Or to return to the Farm just as the Palstrey barouche turned in at the gate with Lady Walderhurst sitting in it glowing with health and that enjoyable interest in all things which gave her a kind of radiance of eye and colour.

She came at length in a time when she did not look quite so radiant. This, it appeared, was from a reason which might be regarded as natural under the circumstances. A more ardent man than Lord Walderhurst might have felt that he could not undertake a journey to foreign lands which would separate him from a wife comparatively new. But Lord Walderhurst was not ardent, and he had married a woman who felt that he did all things well—that, in fact, a thing must be well because it was his choice to do it. His journey to India might, it was true, be a matter of a few months, and involved diplomatic business for which a certain unimpeachable respectability was required. A more brilliant man, who had been less respectable in the most decorous British sense, would not have served the purpose of the government.

Emily's skin had lost a shade of its healthful freshness, it struck Hester, when she saw her. There was a suggestion of fulness under her eyes. Yet with the bright patience of her smile she defied the remote suspicion that she had shed a tear or so before leaving home. She explained the situation with an affectionally reverent dwelling upon the dignity of the mission which would temporarily bereave her of her mate. Her belief in Walderhurst's intellectual importance to the welfare of the government was a complete and touching thing.

"It will not be for very long," she said, "and you and I must see a great deal of each other. I am so glad you are here. You know how one misses—" breaking off with an admirable air of determined cheer—"I must not think of that."

Walderhurst congratulated himself seriously during the days before his departure. She was so exactly what he liked a woman to be. She might have made difficulties, or have been sentimental. If she had been a girl, it would have been necessary to set up a sort of nursery for her, but this fine amenable, sensible creature could take perfect care of herself. It was only necessary to express a wish, and she not only knew how to carry it out, but was ready to do so without question. As far as he was concerned, he was willing to leave all to her own taste. It was such decent taste. She had no modern ideas which might lead during his absence to any action likely to disturb or annoy him. What she would like best to do would be to stay at Palstrey and enjoy the beauty of it. She would spend her days in strolling through the gardens, talking to the gardeners, who had all grown fond of her, or paying little visits to old people or young ones in the village. She would help the vicar's wife in her charities, she would appear in the Manor pew at church regularly, make the necessary dull calls, and go to the unavoidable dull dinners with a faultless amiability and decorum.

"As I remarked when you told me you had asked her to marry you," said Lady Maria on the occasion of his lunching with her on running up to town for a day's business, "you showed a great deal more sense than most men of your age and rank. If people *will* marry, they should choose the persons least likely to interfere with them. Emily will never interfere with you. She cares a great deal more about your pleasure than her own. And as to that, she's so much like a big, healthy, good child that she would find pleasure wheresoever you dropped her."

This was true, yet the healthy, childish creature had, in deep privacy, cried a little, and was pathetically glad to feel that the Osborns were to be near her, and that she would have Hester to think of and take care of during the summer.

It was pathetic that she should cherish an affection so ingenuous for the Osborns, for one of them at least had no patience with her. To Captain Osborn her existence and presence in the near neighbourhood were offences. He told himself that she was of the particular type of woman he most disliked. She was a big, blundering fool, he said, and her size and very good nature itself got on his nerves and irritated him.

"She looks so deucedly prosperous with her first-rate clothes and her bouncing health," he said.

"The tread of her big feet makes me mad when I hear it."

Hester answered with a shrill little laugh.

"Her big feet are a better shape than mine," she said. "I ought to hate her, and I would if I could, but I can't."

"I can," muttered Osborn between his teeth as he turned to the mantel and scratched a match to light his pipe.

CHAPTER TWELVE

When Lord Walderhurst took his departure for India, his wife began to order her daily existence as he had imagined she would. Before he had left her she had appeared at the first Drawing-room, and had spent a few weeks at the town house, where they had given several imposing and serious dinner parties, more remarkable for dignity and good taste than liveliness. The duties of social existence in town would have been unbearable for Emily without her husband. Dressed by Jane Cupp with a passion of fervour, fine folds sweeping from her small, long waist, diamonds strung round her neck, and a tiara or a big star in her full brown hair, Emily was rather superb when supported by the consciousness that Walderhurst's well-carried maturity and long accustomedness were near her. With him she could enjoy even the unlively splendour of a function, but without him she would have been very unhappy. At Palstrey she was ceasing to feel new, and had begun to realise that she belonged to the world she lived in. She was becoming accustomed to her surroundings, and enjoyed them to the utmost. Her easily roused affections were warmed by the patriarchal atmosphere of village life. Most of the Palstrey villagers had touched their forelocks or curtsied to Walderhursts for generations. Emily liked to remember this, and had at once conceived a fondness for the simple folk, who seemed somehow related so closely to the man she worshipped.

Walderhurst had not the faintest conception of what this worship represented. He did not even reach the length of realising its existence. He saw her ingenuous reverence for and belief in him, and was naturally rather pleased by them. He was also vaguely aware that if she had been a more brilliant woman she would have been a more exacting one, and less easily impressed. If she had been a stupid woman or a clumsy one, he would have detested her and bitterly regretted his marriage. But she was only innocent and gratefully admiring, which qualities, combining themselves with good looks, good health, and good manners, made of a woman something he liked immensely. Really she had looked very nice and attractive when she had bidden him good-by, with her emotional flush and softness of expression and the dewy brightness of her eyes. There was something actually moving in the way her strong hand had wrung his at the last moment.

"I only *wish*," she had said, "I only do so *wish* that there was something I could *do* for you while you are away—something you could leave me to *do*."

"Keep well and enjoy yourself," he had answered. "That will really please me."

Nature had not so built him that he could suspect that she went home and spent the rest of the morning in his rooms, putting away his belongings with her own hands, just for the mere passion of comfort she felt in touching the things he had worn, the books he had handled, the cushions his head had rested against. She had indeed mentioned to the housekeeper at Berkeley Square that she wished his lordship's apartments to remain untouched until she herself had looked over them. The obsession which is called Love is an emotion past all explanation. The persons susceptible to its power are as things beneath a spell. They see, hear, and feel that of which the rest of their world is unaware, and will remain unaware for ever. To the endearing and passion-inspiring qualities Emily Walderhurst saw in this more than middle-aged gentleman an unstirred world would remain blind, deaf, and imperceptive until its end transpired. This, however, made not the slightest difference in the reality of these things as she saw and felt and was moved to her soul's centre by them. Bright youth in Agatha Norman, at present joyously girdling the globe with her bridegroom, was moved much less deeply, despite its laughter and love.

A large lump swelled in Emily's throat as she walked about the comfortable, deserted apartments of her James. Large tears dropped on the breast of her dress as they had dropped upon her linen blouse when she walked across the moor to Maundell. But she bravely smiled as she tenderly brushed away with her hand two drops which fell upon a tweed waistcoat she had picked up. Having done this, she suddenly stooped and kissed the rough cloth fervently, burying her face in it with a sob.

"I do *love* him so!" she whispered, hysterically. "I do so *love* him, and I shall so *miss* him!" with the italicised feelingness of old.

The outburst was in fact so strongly italicised that she felt the next moment almost as if she had been a little indecent. She had never been called upon by the strenuousness of any occasion to mention baldly to Lord Walderhurst that she "loved" him. It had not been necessary, and she was too little used to it not to be abashed by finding herself proclaiming the fact to his very waistcoat itself. She sat down holding the garment in her hands and let her tears fall.

She looked about her at the room and across the corridor through the open door at his study which adjoined it. They were fine rooms, and every book and bust and chair looked singularly suggestive of his personality. The whole house was beautiful and imposing in Emily's eyes. "He has made all my life beautiful and full of comfort and happiness," she said, trembling. "He has saved me from everything I was afraid of, and there is nothing I can *do*. Oh!" suddenly dropping a hot face on her hands, "if I were only Hester Osborn. I should be glad to suffer anything, or die in any way. I should have paid him back—just a little—if I might."

For there was one thing she had learned through her yearning fervour, not through any speech of his. All the desire and pride in him would be fed full and satisfied if he could pass his name on to a creature of his own flesh and blood. All the heat his cold nature held had concentrated itself in a secret passion centred on this thing. She had begun to awaken to a suspicion of this early in their marriage, and afterwards by processes of inclusion and exclusion she had realised the proud intensity of his feeling despite his reserve and silence. As for her, she would have gone to the stake, or have allowed her flesh to be cut into pieces to form that which would have given him reason for exultation and pride. Such was the helpless, tragic, kindly love and yearning of her.

* * * * *

The thing filled her with a passion of tenderness for Hester Osborn. She yearned over her, too. Her spinster life had never brought her near to the mystery of birth. She was very ignorant and deeply awed by the mere thought of it. At the outset Hester had been coldly shy and reticent, but as they saw each other more she began to melt before the unselfish warmth of the other woman's overtures of friendship. She was very lonely and totally inexperienced. As Agatha Slade had gradually fallen into intimacy of speech, so did she. She longed so desperately for companionship that the very intensity of her feelings impelled her to greater openness than she had at first intended.

"I suppose men don't know," she said to herself sullenly, in thinking of Osborn, who spent his days out of doors. "At any rate, they don't care."

Emily cared greatly, and was so full of interest and sympathy that there was something like physical relief in talking to her.

"You two have become great pals," Alec said, on an afternoon when he stood at a window watching Lady Walderhurst's carriage drive away. "You spend hours together talking. What is it all about?"

"She talks a good deal about her husband. It is a comfort to her to find someone to listen. She thinks he is a god. But we principally talk about—me."

"Don't discourage her," laughed Osborn. "Perhaps she will get so fond of you that she will not be willing to part with us, as she will be obliged to take both to keep one."

"I wish she would, I wish she would!" sighed Hester, tossing up her hands in a languid, yet fretted gesture.

The contrast between herself and this woman was very often too great to be equably borne. Even her kindness could not palliate it. The simple perfection of her country clothes, the shining skins of her horses, the smooth roll of her carriage, the automatic servants who attended her, were suggestive of that ease and completeness in all things, only to be compassed by long-possessed wealth. To see every day the evidences of it while one lived on charitable sufferance on the crumbs which fell from the master's table was a galling enough thing, after all. It would always have been galling. But it mattered so much more now—so much more to Hester than she had known it could matter even in those days when as a girl she had thirstily longed for it. In those days she had not lived near enough to it all to know the full meaning and value of it—the beauty and luxury, the stateliness and good taste. To have known it in this way, to have been almost part of it and then to leave it, to go back to a hugger-mugger existence in a wretched bungalow hounded by debt, pinched and bound hard and fast by poverty, which offered no future prospect of bettering itself into decent good luck! Who could bear it?

Both were thinking the same thing as their eyes met.

"How are we to stand it, after this?" she cried out sharply.

"We can't stand it," he answered. "Confound it all, something *must* happen."

"Nothing will," she said; "nothing but that we shall go back worse off than before."

* * * * *

At this period Lady Walderhurst went to London again to shop, and spent two entire happy days in buying beautiful things of various kinds, which were all to be sent to Mrs. Osborn at The Kennel Farm, Palstrey. She had never enjoyed herself so much in her life as she did during those two days when she sat for hours at one counter after another looking at exquisite linen and flannel and lace. The days she had spent with Lady Maria in purchasing her trousseau had not compared with these two. She looked actually lovely as she almost fondled the fine fabrics, smiling with warm softness at the pretty things shown her. She spent, in fact, good deal of money, and luxuriated in so doing as she tfould never have luxuriated in spending it in finery for herself. Nothing indeed seemed too fairy-like in its fineness, no quantity of lace seemed in excess. Her heart positively trembled in her breast sometimes, and she found strange tears rising in her eyes.

"They are so sweet," she said plaintively to the silence of her own bedroom as she looked some of her purchases over. "I don't know why they give me such a feeling. They look so little and— helpless, and as if they were made to hold in one's arms. It's absurd of me, I daresay."

The morning the boxes arrived at The Kennel Farm, Emily came too. She was in the big carriage, and carried with her some special final purchases she wanted to bring herself. She came because she could not have kept away. She wanted to see the things again, to be with Hester when she unpacked them, to help her, to look them all over, to touch them and hold them in her hands.

She found Hester in the large, low-ceilinged room in which she slept. The big four-post bed was already snowed over with a heaped-up drift of whiteness, and open boxes were scattered about. There was an odd expression in the girl's eyes, and she had a red spot on either cheek.

"I did not expect anything like this," she said. "I thought I should have to make some plain, little things myself, suited to its station," with a wry smile. "They would have been very ugly. I don't know how to sew in the least. You forget that you were not buying things for a prince or a princess, but for a little beggar."

"Oh, don't!" cried Emily, taking both her hands. "Let us be *happy*! It was so *nice* to buy them. I never liked anything so much in my life."

She went and stood by the bedside, taking up the things one by one, touching up frills of lace and smoothing out tucks.

"Doesn't it make you happy to look at them?" she said.

"*You* look at them," said Hester, staring at her, "as if the sight of them made you hungry, or as if you had bought them for yourself."

Emily turned slightly away. She said nothing. For a few moments there was a dead silence.

Hester spoke again. What in the world was it in the mere look of the tall, straight body of the woman to make her feel hot and angered.

"If you had bought them for yourself," she persisted, "they would be worn by a Marquis of Walderhurst."

Emily laid down the robe she had been holding. She put it on the bed, and turned round to look at Hester Osborn with serious eyes.

"They *may* be worn by a Marquis of Walderhurst, you know," she answered. "They may."

She was remotely hurt and startled, because she felt in the young woman something she had felt once or twice before, something resentful in her thoughts of herself, as if for the moment she represented to her an enemy.

The next moment, however, Hester Osborn fell upon her with embraces.

"You are an angel to me," she cried. "You are an angel, and I can't thank you. I don't know how."

Emily Walderhurst patted her shoulder as she kindly enfolded her in warm arms.

"Don't thank me," she half whispered emotionally. "Don't. Just let us *enjoy* ourselves."

CHAPTER THIRTEEN

Alec Osborn rode a good deal in these days. He also walked a good deal, sometimes with a gun over his shoulder and followed by a keeper, sometimes alone. There was scarcely a square yard of the Palstrey Manor lands he had not tramped over. He had learned the whole estate by heart, its woods, its farms, its moorlands. A morbid secret interest in its beauties and resources possessed him. He could not resist the temptation to ask apparently casual questions of keepers and farmers when he found himself with them. He managed to give his inquiries as much the air of accident as possible, but he himself knew that they were made as a result of a certain fevered curiosity. He found that he had fallen into the habit of continually making plans connected with the place. He said to himself, "If it were mine I would do this, or that. If I owned it, I would make this change or that one. I would discharge this keeper or put another man on such a farm." He tramped among the heather thinking these things over, and realising to the full what the pleasure of such powers would mean to a man such as himself, a man whose vanity had never been fed, who had a desire to control and a longing for active out-of-door life.

"If it were mine, if it were mine!" he would say to himself. "Oh! damn it all, if it were only mine!"

And there were other places as fine, and finer places he had never seen,—Oswyth, Hurst, and Towers,—all Walderhurst's all belonging to this one respectable, elderly muff. Thus he summed up the character of his relative. As for himself he was young, strong, and with veins swelling with the insistent longing for joyful, exultant life. The sweating, panting drudgery of existence in India was a thought of hell to him. But there it was, looming up nearer and nearer with every heavenly English day that passed. There was nothing for it but to go back—go back, thrust one's neck into the collar again, and sweat and be galled to the end. He had no ambitions connected with his profession. He realised loathingly in these days that he had always been waiting, waiting.

The big, bright-faced woman who was always hanging about Hester, doing her favours, he actually began to watch feverishly. She was such a fool; she always looked so healthy, and she was specially such a fool over Walderhurst. When she had news of him, it was to be seen shining in her face.

She had a sentimental school-girl fancy that during his absence she would apply herself to the task of learning to ride. She had been intending to do so before he went away; they had indeed spoken of it together, and Walderhurst had given her a handsome, gentle young mare. The creature was as kind as she was beautiful. Osborn, who was celebrated for his horsemanship, had promised to undertake to give the lessons.

A few days after her return from London with her purchases, she asked the husband and wife to lunch with her at Palstrey, and during the meal broached the subject.

"I should like to begin soon, if you can spare the time for me," she said. "I want to be able to go out with him when he comes back. Do you think I shall be slow in learning? Perhaps I ought to be lighter to ride well."

"I think you will be pretty sure to have a first-class seat," Osborn answered. "You will be likely to look particularly well."

"Do you think I shall? How good you are to encourage me. How soon could I begin?"

She was quite agreeably excited. In fact, she was delighted by innocent visions of herself as Walderhurst's equestrian companion. Perhaps if she sat well, and learned fine control of her horse, he might be pleased, and turn to look at her, as they rode side by side, with that look of approval and dawning warmth which brought such secret joy to her soul.

"When may I take my first lesson?" she said quite eagerly to Captain Osborn, for whom a footman was pouring out a glass of wine.

"As soon," he answered, "as I have taken out the mare two or three times myself. I want to know her thoroughly. I would not let you mount her until I had learned her by heart."

They went out to the stables after lunch and visited the mare in her loose box. She was a fine beast, and seemed as gentle as a child.

Captain Osborn asked questions of the head groom concerning her. She had a perfect reputation, but nevertheless she was to be taken over to the Kennel stables a few days before Lady Walderhurst mounted her.

"It is necessary to be more than careful," Osborn said to Hester that night. "There would be the devil and all to pay if anything went wrong."

The mare was brought over the next morning. She was a shining bay, and her name was Faustine.

In the afternoon Captain Osborn took her out. He rode her far and learned her thoroughly before he brought her back. She was as lively as a kitten, but as kind as a dove. Nothing could have been better tempered and safer. She would pass anything, even the unexpected appearance of a road-mending engine turning a corner did not perceptibly disturb her.

"Is she well behaved?" Hester asked at dinner time.

"Yes, apparently," was his answer; "but I shall take her out once or twice again."

He did take her out again, and had only praise for her on each occasion. But the riding lessons did not begin at once. In fact he was, for a number of reasons, in a sullen and unsociable humour which did not incline him towards the task he had undertaken. He made various excuses for not beginning the lessons, and took Faustine out almost every day.

But Hester had an idea that he did not enjoy his rides. He used to return from them with a resentful, sombre look, as if his reflections had not been pleasant company for him. In truth they were not pleasant company. He was beset by thoughts he did not exactly care to be beset by—thoughts which led him farther than he really cared to go, which did not incline him to the close companionship of Lady Walderhurst. It was these thoughts which led him on his long rides; it was one of them which impelled him, one morning, as he was passing a heap of broken stone, piled for the mending of the ways by the roadside, to touch Faustine with heel and whip. The astonished young animal sprang aside curvetting. She did not understand, and to horse-nature the uncomprehended is alarming. She was more bewildered and also more fretted when, in passing the next stone heap, she felt the same stinging touches. What did it mean? Was she to avoid this thing, to leap at sight of it, to do what? She tossed her delicate head and snorted in her trouble. The country road was at some distance from Palstrey, and was little frequented. No one was in sight. Osborn glanced about him to make sure of this fact. A long stretch of road lay before him, with stone heaps piled at regular intervals. He had taken a big whiskey and soda at the last wayside inn he had passed, and drink did not make him drunk so much as mad. He pushed the mare ahead, feeling in just the humour to try experiments with her.

* * * * *

"Alec is very determined that you shall be safe on Faustine," Hester said to Emily. "He takes her out every day."

"It is very good of him," answered Emily.

Hester thought she looked a trifle nervous, and wondered why. She did not say anything about the riding lessons, and in fact had seemed of late less eager and interested. In the first place, it had been Alec who had postponed, now it was she. First one trifling thing and then another seemed to interpose.

"The mare is as safe as a feather-bed," Osborn said to her one afternoon when they were taking tea on the lawn at Palstrey. "You had better begin now if you wish to accomplish anything before Lord Walderhurst comes back. What do you hear from him as to his return?"

Emily had heard that he was likely to be detained longer than he had expected. It seemed always to be the case that people were detained by such business. He was annoyed, but it could not be helped. There was a rather tired look in her eyes and she was paler than usual.

"I am going up to town to-morrow," she said. "The riding lessons might begin after I come back."

"Are you anxious about anything?" Hester asked her as she was preparing for the drive back to The Kennel Farm.

"No, no," Emily answered. "Only——"

"Only what?"

"I should be so glad if—if he were not away."

Hester gazed reflectively at her suddenly quivering face.

"I don't think I ever saw a woman so fond of a man," she said.

Emily stood still. She was quite silent. Her eyes slowly filled. She had never been able to say much about what she felt for Walderhurst. Hers was a large, dumb, primitive affection.

She sat at her open bedroom window a long time that evening. She rested her chin upon her hand and looked up at the deeps of blue powdered with the diamond dust of stars. It seemed to her that she had never looked up and seen such myriads of stars before. She felt far away from earthly things and tremulously uplifted. During the last two weeks she had lived in a tumult of mind, of amazement, of awe, of hope and fear. No wonder that she looked pale and that her face was full of anxious yearning. There were such wonders in the world, and she, Emily Fox-Seton, no, Emily Walderhurst, seemed to have become part of them.

She clasped her hands tight together and leaned forward into the night with her face turned upwards. Very large drops began to roll fast down her cheeks, one after the other. The argument of scientific observation might have said she was hysterical, and whether with or without reason is immaterial. She did not try to check her tears or wipe them away, because she did not know that she was crying. She began to pray, and heard herself saying the Lord's Prayer like a child.

"Our Father who art in Heaven—Our Father who art in Heaven, hallowed be Thy name," she murmured imploringly.

She said the prayer to the end, and then began it over again. She said it three or four times, and her appeal for daily bread and the forgiveness of trespasses expressed what her inarticulate nature could not have put into words. Beneath the entire vault of heaven's dark blue that night there was nowhere lifted to the Unknown a prayer more humbly passion-full and gratefully imploring than her final whisper.

"For Thine is the kingdom, the power, and the glory, for ever and ever. Amen, amen."

When she left her seat at the window and turned towards the room again, Jane Cupp, who was preparing for the morrow's journey and was just entering with a dress over her arm, found herself restraining a start at sight of her.

"I hope you are quite well, my lady," she faltered.

"Yes," Lady Walderhurst answered. "I think I am very well—very well, Jane. You will be quite ready for the early train to-morrow morning."

"Yes, my lady, quite."

"I have been thinking," said Emily gently, almost in a tone of reverie, "that if your uncle had not wanted your mother so much it would have been nice to have her here with us. She is such an experienced person, and so kind. I never forget how kind she was to me when I had the little room in Mortimer Street."

"Oh! my lady, you was kind to *us*," cried Jane.

She recalled afterwards, with tears, how her ladyship moved nearer to her and took her hand with what Jane called "her wonderful *good* look," which always brought a lump to her throat.

"But I always count on you, Jane," she said. "I count on you so much."

"Oh! my lady," Jane cried again, "it's my comfort to believe it. I'd lay down my life for your ladyship, I would indeed."

Emily sat down, and on her face there was a soft, uplifted smile.

"Yes," she said, and Jane Cupp saw that she was reflective again, and the words were not addressed exactly, to herself, "one would be quite ready to lay down one's life for the person one loved. It seems even a little thing, doesn't it?"

CHAPTER FOURTEEN

Lady Walderhurst remained in town a week, and Jane Cupp remained with her, in the house in Berkeley Square, which threw open its doors to receive them on their arrival quite as if they had never left it. The servants' hall brightened temporarily in its hope that livelier doings might begin to stir the establishment, but Jane Cupp was able to inform inquirers that the visit was only to be a brief one.

"We are going back to Palstrey next Monday," she explained. "My lady prefers the country, and she is very fond of Palstrey; and no wonder. It doesn't seem at all likely she'll come to stay in London until his lordship gets back."

"We hear," said the head housemaid, "that her ladyship is very kind to Captain Osborn and his wife, and that Mrs. Osborn's in a delicate state of health."

"It would be a fine thing for us if it was in our family," remarked an under housemaid who was pert.

Jane Cupp looked extremely reserved.

"Is it true," the pert housemaid persisted, "that the Osborns can't abide her?"

"It's true," said Jane, severely, "that she's goodness itself to them, and they ought to adore her."

"We hear they don't," put in the tallest footman. "And who wonders. If she was an angel, there's just a chance that she may give Captain Osborn a wipe in the eye, though she is in her thirties."

"It's not for *us*," said Jane, stiffly, "to discuss thirties or forties or fifties either, which are no business of ours. There's one gentleman, and him a marquis, as chose her over the heads of two beauties in their teens, at least."

"Well, for the matter of that," admitted the tall footman, "I'd have chose her myself, for she's a fine woman."

Lady Maria was just on the point of leaving South Audley Street to make some visits in the North, but she came and lunched with Emily, and was in great form.

She had her own opinion of a number of matters, some of which she discussed, some of which she kept to herself. She lifted her gold lorgnette and looked Emily well over.

"Upon my word, Emily," she said, "I am proud of you. You are one of my successes. Your looks are actually improving. There's something rather etherealised about your face to-day. I quite agree with Walderhurst in all the sentimental things he says about you."

She said this last partly because she liked Emily and knew it would please her to hear that her husband went to the length of dwelling on her charms in his conversation with other people, partly because it entertained her to see the large creature's eyelids flutter and a big blush sweep her cheek.

"He really was in great luck when he discovered you," her ladyship went on briskly. "As for that, I was in luck myself. Suppose you had been a girl who could not have been left. As Walderhurst is short of female relatives, it would have fallen to me to decently dry-nurse you. And there would have been the complications arising from a girl being baby enough to want to dance about to places, and married enough to feel herself entitled to defy her chaperone; she couldn't have been trusted to chaperone herself. As it is, Walderhurst, can go where duty calls, etc., and I can make my visits and run about, and you, dear thing, are quite happy at Palstrey playing Lady Bountiful and helping the little half-breed woman to expect her baby. I daresay you sit and make dolly shirts and christening robes hand in hand."

"We enjoy it all very much," Emily answered, adding imploringly, "please don't call her a little half-breed woman. She's such a dear little thing, Lady Maria."

Lady Maria indulged in the familiar chuckle and put up her lorgnette to examine her again.

"There's a certain kind of early Victorian saintliness about you, Emily Walderhurst, which makes my joy," she said. "You remind me of Lady Castlewood, Helen Pendennis, and Amelia Sedley, with the spitefulness and priggishness and catty ways left out. You are as nice as Thackeray *thought* they were, poor mistaken man. I am not going to suffuse you with blushes by explaining to you that there is what my nephew would call a jolly good reason why, if you were not an early Victorian and improved Thackerayian saint, you would not be best pleased at finding yourself called upon to assist at this interesting occasion. Another kind of woman would probably feel like a cat towards the little Osborn. But even the mere reason itself, as a reason, has not once risen in your benign and pellucid mind. You have a pellucid mind, Emily; I should be rather proud of the word if I had invented it myself to describe you. But I didn't. It was Walderhurst. You have actually wakened up the man's intellects, such as they are."

She evidently had a number of opinions of the Osborns. She liked neither of them, but it was Captain Osborn she especially *dis*liked.

"He is really an underbred person," she explained, "and he hasn't the sharpness to know that is the reason Walderhurst detests him. He had vulgar, cheap sort of affairs, and nearly got into the kind of trouble people don't forgive. What a fool a creature in his position is to offend the taste of the man he may inherit from, and who, if he were not antagonistic to him, would regard him as a sort of duty. It wasn't his immorality particularly. Nobody is either moral or immoral in these days, but penniless persons must be decent. It's all a matter of taste and manners. I haven't any morals myself, my dear, but I have beautiful manners. A woman can have the kind of manners which keep her from breaking the Commandments. As to the Commandments, they are awfully easy things *not* to break. Who wants to break them, good Lord! Thou shall do no murder. Thou shalt not steal. Thou shalt not commit, etc. Thou shalt not bear false witness. That's simply gossip and lying, and they are bad manners. If you have good manners, you *don't*."

She chatted on in her pungent little worldly, good-humoured way through the making of a very excellent lunch. After which she settled her smart bonnet with clever touches, kissed Emily on both cheeks, and getting into her brougham rolled off smiling and nodding.

Emily stood at the drawing-room window and watched her equipage roll round the square and into Charles Street, and then turned away into the big, stately empty room, sighing without intending to do so while she smiled herself.

"She's so witty and so amusing," she said; "but one would no more think of *telling* her anything than one would think of catching a butterfly and holding it while one made it listen. She would be so *bored* if she was confided in."

Which was most true. Never in her life had her ladyship allowed herself the indiscretion of appearing a person in whom confidences might be reposed. She had always had confidences enough of her own to take care of, without sharing those of other people.

"Good heavens!" she had exclaimed once, "I should as soon think of assuming another woman's wrinkles."

On the first visit Lady Walderhurst made to The Kennel Farm the morning after her return to Palstrey, when Alec Osborn helped her from her carriage, he was not elated by the fact that he had never seen her look so beautifully alive and blooming during his knowledge of her. There was a fine rose on her cheek, and her eyes were large and happily illumined.

"How well you look!" broke from him with an involuntariness he was alarmed to realise as almost spiteful. The words were an actual exclamation which he had not meant to utter, and Emily Walderhurst even started a trifle and looked at him with a moment's question.

"But you look well, too," she answered. "Palstrey agrees with both of us. You have such a colour."

"I have been riding," he replied. "I told you I meant to know Faustine thoroughly before I let you mount her. She is ready for you now. Can you take your first lesson to-morrow?"

"I—I don't quite know," she hesitated. "I will tell you a little later. Where is Hester?"

Hester was in the drawing-room. She was lying on a sofa before an open window and looking rather haggard and miserable. She had, in fact, just had a curious talk with Alec which had ended in something like a scene. As Hester's health grew more frail, her temper became more fierce, and of late there had been times when a certain savagery, concealed with difficulty in her husband's moods, affected her horribly.

This morning she felt a new character in Emily's manner. She was timid and shy, and a little awkward. Her child-like openness of speech and humour seemed obscured. She had less to say than usual, and at the same time there was a suggestion of restless unease about her. Hester Osborn, after a few minutes, began to have an odd feeling that the woman's eyes held a question or a desire in them.

She had brought some superb roses from the Manor gardens, and she moved about arranging them for Hester in vases.

"It is beautiful to come back to the country," she said. "When I get into the carriage at the station and drive through the sweet air, I always feel as if I were beginning to live again, and as if in London I had not been quite alive. It seemed so *heavenly* in the rose garden at Palstrey to-day, to walk about among those thousands of blooming lovely things breathing scent and nodding their heavy, darling heads."

"The roads are in a beautiful condition for riding," Hester said, "and Alec says that Faustine is perfect. You ought to begin to-morrow morning. Shall you?"

She spoke the words somewhat slowly, and her face did not look happy. But, then, it never was a really happy face. The days of her youth had been too full of the ironies of disappointment.

There was a second's silence, and then she said again:

"Shall you, if it continues fine?"

Emily's hands were full of roses, both hands, and Hester saw both hands and roses tremble. She turned round slowly and came towards her. She looked nervous, awkward, abashed, and as if for that moment she was a big girl of sixteen appealing to her and overwhelmed with queer feelings, and yet the depths of her eyes held a kind of trembling, ecstatic light. She came and stood before her, holding the trembling roses as if she had been called up for confession.

"I—I mustn't," she half whispered. The corners of her lips drooped and quivered, and her voice was so low that Hester could scarcely hear it. But she started and half sat up.

"You *mustn't*?" she gasped; yes, really it was gasped.

Emily's hand trembled so that the roses began to fall one by one, scattering a rain of petals as they dropped.

"I mustn't," she repeated, low and shakily. "I had—reason.—I went to town to see—somebody. I saw Sir Samuel Brent, and he told me I must not. He is quite sure."

She tried to calm herself and smile. But the smile quivered and ended in a pathetic contortion of her face. In the hope of gaining decent self-control, she bent down to pick up the dropped roses. Before she had picked up two, she let all the rest fall, and sank kneeling among them, her face in her hands.

"Oh, Hester, Hester!" she panted, with sweet, stupid unconciousness of the other woman's heaving chest and glaring eyes. "It has come to me too, actually, after all."

CHAPTER FIFTEEN

The Palstrey Manor carriage had just rolled away carrying Lady Walderhurst home. The big, low-ceilinged, oak-beamed farmhouse parlour was full of the deep golden sunlight of the late afternoon, the air was heavy with the scent of roses and sweet-peas and mignonette, the adorable fragrance of English country-house rooms. Captain Osborn inhaled it at each breath as he stood and looked out of the diamond-paned window, watching the landau out of sight. He felt the scent and the golden glow of the sunset light as intensely as he felt the dead silence which reigned between himself and Hester almost with the effect of a physical presence. Hester was lying upon the sofa again, and he knew she was staring at his back with that sardonic widening of her long eyes, a thing he hated, and which always foreboded things not pleasant to face.

He did not turn to face them until the footman's cockade had disappeared finally behind the tall hedge, and the tramp of the horses' feet was deadening itself in the lane. When he ceased watching and listening, he wheeled round suddenly.

"What does it all mean?" he demanded. "Hang her foolish airs and graces. *Why* won't she ride, for she evidently does not intend to."

Hester laughed, a hard, short, savage little un-mirthful sound it was.

"No, she doesn't intend to," she answered, "for many a long day, at least, for many a month. She has Sir Samuel Brent's orders to take the greatest care of herself."

"Brent's? Brent's?"

Hester struck her lean little hands together and laughed this time with a hint at hysteric shrillness.

"I told you so, I told you so!" she cried. "I knew it would be so, I knew it! By the time she reaches her thirty-sixth birthday there will be a new Marquis of Walderhurst, and he won't be either you or yours." And as she finished, she rolled over on the sofa, and bit the cushions with her teeth as she lay face downwards on them. "He won't be you, or belong to you," she reiterated, and then she struck the cushions with her clenched fist.

He rushed over to her, and seizing her by the shoulders shook her to and fro.

"You don't know what you are talking about," he said; "you don't know what you are saying."

"I do! I do! I do!" she screamed under her breath, and beat the cushions at every word. "It's true, it's true. She's drivelling about it, drivelling!"

Alec Osborn threw back his head, drawing in a hard breath which was almost a snort of fury.

"By God!" he cried, "if she went out on Faustine now, she would not come back!"

His rage had made him so far beside himself that he had said more than he intended, far more than he would have felt safe. But the girl was as far beside herself as he was, and she took him up.

"Serve her right," she cried. "I shouldn't care. I hate her! I hate her! I told you once I couldn't, but I do. She's the biggest fool that ever lived. She knew *nothing* of what I felt. I believe she thought I would rejoice with her. I didn't know whether I should shriek in her face or scream out laughing. Her eyes were as big as saucers, and she looked at me as if she felt like the Virgin Mary after the Annunciation. Oh! the stupid, *inhuman* fool!"

Her words rushed forth faster and faster, she caught her breath with gasps, and her voice grew more shrill at every sentence. Osborn shook her again.

"Keep quiet," he ordered her. "You are going into hysterics, and it won't do. Get hold of yourself."

"Go for Ameerah," she gasped, "or I'm afraid I can't. She knows what to do."

He went for Ameerah, and the silently gliding creature came bringing her remedies with her. She looked at her mistress with stealthily questioning but affectionate eyes, and sat down on the floor rubbing her hands and feet in a sort of soothing massage. Osborn went out of the room, and the two women were left together. Ameerah knew many ways of calming her mistress's nerves, and perhaps one of the chief ones was to lead her by subtle powers to talk out her rages and anxieties. Hester never knew that she was revealing herself and her moods until after her interviews with the Ayah were over. Sometimes an hour or so had passed before she began to realise that she had let out things which she had meant to keep secret. It was never Ameerah who talked, and Hester was never conscious that she talked very much herself. But afterwards she saw that the few sentences she had uttered were such as would satisfy curiosity if the Ayah felt it. Also she was not, on the whole, at all sure that the woman felt it. She showed no outward sign of any interest other than the interest of a deep affection. She loved her young mistress to-day as passionately as she had loved her as a child when she had held her in her bosom as if she had been her own. By the time Emily Walderhurst had reached Palstrey, Ameerah knew many things. She understood that her mistress was as one who, standing upon the brink of a precipice, was being slowly but surely pushed over its edge—pushed, pushed by Fate. This was the thing imaged in her mind when she shut herself up in her room and stood alone in the midst of the chamber clenching her dark hands high above her white veiled head, and uttering curses which were spells, and spells which were curses.

Emily was glad that she had elected to be alone as much as possible, and had not invited people to come and stay with her. She had not invited people, in honest truth, because she felt shy of the responsibility of entertainment while Walderhurst was not with her. It would have been proper to invite his friends, and his friends were all people she was too much in awe of, and too desirous to please to be able to enjoy frankly as society. She had told herself that when she had been married a few years she would be braver.

And now her gladness was so devout that it was pure rejoicing. How could she have been calm, how could she have been conversational, while through her whole being there surged but one thought. She was sure that while she talked to people she would have been guilty of looking as if she was thinking of something not in the least connected with themselves.

If she had been less romantically sentimental in her desire to avoid all semblance of burdening her husband she would have ordered him home at once, and demanded as a right the protection of his dignity and presence. If she had been less humble she would have felt the importance of her position and the gravity of the claims it gave her to his consideration, instead of being lost in prayerful gratitude to heaven.

She had been rather stupidly mistaken in not making a confidante of Lady Maria Bayne, but she had been, in her big girl shyness, entirely like herself. In some remote part of her nature she had shrunk from a certain look of delighted amusement which she had known would have betrayed itself, despite her ladyship's good intentions, in the eyes assisted by the smart gold lorgnette. She knew she was inclined to be hyper-emotional on this subject, and she felt that if she had seen the humour trying to conceal itself behind the eye-glasses, she might have been hysterical enough to cry even while she tried to laugh, and pass her feeling off lightly. Oh, no! Oh, no! Somehow she *knew* that at such a moment, for some fantastic, if subtle, reason, Lady Maria would only see her as Emily Fox-Seton, that she would have actually figured before her for an instant as poor Emily Fox-Seton making an odd confession. She could not have endured it without doing something foolish, she felt that she would not, indeed.

So Lady Maria went gaily away to make her round of visits and be the amusing old life and soul of house-party after house-party, suspecting nothing of a possibility which would actually have sobered her for a moment.

Emily passed her days at Palstrey in a state of happy exaltation. For a week or so they were spent in wondering whether or not she should write a letter to Lord Walderhurst which should convey the information to him which even Lady Maria would have regarded as important, but the more she argued the question with herself, the less she wavered from her first intention. Lady Maria's frank congratulation of herself and Lord Walderhurst in his wife's entire unexactingness had indeed been the outcome of a half-formed intention to dissipate amiably even the vaguest inclination to verge on expecting things from people. While she thought Emily unlikely to allow herself to deteriorate into an encumbrance, her ladyship had seen women in her position before, whose marriages had made perfect fools of them through causing them to lose their heads completely and require concessions and attentions from their newly acquired relations which bored everybody. So she had lightly patted and praised Emily for the course of action she preferred to "keep her up to."

"She's the kind of woman ideas sink into if they are well put," she had remarked in times gone by. "She's not sharp enough to see that things are being suggested to her, but a suggestion acts upon her delightfully."

Her suggestions acted upon Emily as she walked about the gardens at Palstrey, pondering in the sunshine and soothed by the flower scents of the warmed borders. Such a letter written to Walderhurst might change his cherished plans, concerning which she knew he held certain ambitions. He had been so far absorbed in them that he had gone to India at a time of the year which was not usually chosen for the journey. He had become further interested and absorbed after he had reached the country, and he was evidently likely to prolong his stay as he had not thought of prolonging it. He wrote regularly though not frequently, and Emily had gathered from the tone of his letters that he was more interested than he had ever been in his life before.

"I would not interfere with his work for anything in the world," she said. "He cares more for it than he usually cares for things. I care for everything—I have that kind of mind; an intellectual person is different. I am perfectly well and happy here. And it will be so nice to look forward."

She was not aware how Lady Maria's suggestions had "sunk in." She would probably have reached the same conclusion without their having been made, but since they had been made, they had assisted her. There was one thing of all others she felt she could not possibly bear, which was to realise that she herself could bring to her James's face an expression she had once or twice seen others bring there (Captain Osborn notably),—an expression of silent boredom on the verge of irritation. Even radiant domestic joy might not be able to overrule this, if just at this particular juncture he found himself placed in the position of a man whom decency compelled to take the next steamer to England.

If she had felt tenderly towards Hester Osborn before, the feeling was now increased tenfold. She went to see her oftener, she began to try to persuade her to come and stay at Palstrey. She was all the more kind because Hester seemed less well, and was in desperate ill spirits. Her small face had grown thin and yellow, she had dark rings under her eyes, and her little hands were hot and looked like bird's claws. She did not sleep and had lost her appetite.

"You must come and stay at Palstrey for a few days," Emily said to her. "The mere change from one house to another may make you sleep better."

But Hester was not inclined to avail herself of the invitation. She made obstacles and delayed acceptance for one reason and another. She was, in fact, all the more reluctant because her husband wished her to make the visit. Their opposed opinions had resulted in one of their scenes.

"I won't go," she had said at first. "I tell you I won't."

"You will," he answered. "It will be better for you."

"Will it be worse for me if I don't?" she laughed feverishly. "And how will it be better for you if I do? I know you are in it."

He lost his temper and was indiscreet, as his temper continually betrayed him into being.

"Yes, I am in it," he said through his teeth, "as you might have the sense to see. Everything is the better for us that throws us with them, and makes them familiar with the thought of us and our rights."

"Our rights," the words were a shrill taunt.

"What rights have you, likely to be recognised, unless you kill her. Are you going to kill her?"

He had a moment of insanity.

"I'd kill her and you too if it was safe to do it. You both deserve it!"

He flung across the room, having lost his wits as well as his temper. But a second later both came back to him as in a revulsion of feeling.

"I talk like a melodramatic fool," he cried. "Oh, Hester, forgive me!" He knelt on the floor by her side, caressing her imploringly. "We both take fire in the same way. We are both driven crazy by this damned blow. We're beaten; we may as well own it and take what we can get. She's a fool, but she's better than that pompous, stiff brute Walderhurst, and she has a lot of pull over him he knows nothing about. The smug animal is falling in love with her in his way. She can make him do the decent thing. Let us keep friends with her."

"The decent thing would be a thousand a year," wailed Hester, giving in to his contrition in spite of herself, because she had once been in love with him, and because she was utterly helpless. "Five hundred a year wouldn't be indecent."

"Let us keep on her good side," he said, fondling her, with a relieved countenance. "Tell her you will come and that she is an angel, and that you are sure a visit to the Manor will save your life."

They went to Palstrey a few days later. Ameerah accompanied them in attendance upon her mistress, and the three settled down into a life so regular that it scarcely seemed to wear the aspect of a visit. The Osborns were given some of the most beautiful and convenient rooms in the house. No other visitors were impending and the whole big place was at their disposal. Hester's boudoir overlooked the most perfect nooks of garden, and its sweet chintz draperies and cushions and books and flowers made it a luxurious abode of peace.

"What shall I do," she said on the first evening in it as she sat in a soft chair by the window, looking out at the twilight and talking to Emily. "What shall I do when I must go away?"

"I don't mean only from here,—I mean away from England, to loathly India."

"Do you dislike it so?" Emily asked, roused to a new conception of her feeling by her tone.

"I could never describe to you how much," fiercely. "It is like going to the place which is the opposite of Heaven."

"I did not know that," pityingly. "Perhaps—I wonder if something might not be done: I must talk to my husband."

Ameerah seemed to develop an odd fancy for the society of Jane Cupp, which Jane was obliged to confess to her mistress had a tendency to produce in her system "the creeps."

"You must try to overcome it, Jane," Lady Walderhurst said. "I'm afraid it's because of her colour. I've felt a little silly and shy about her myself, but it isn't nice of us. You ought to read 'Uncle Tom's Cabin,' and all about that poor religious Uncle Tom, and Legree, and Eliza crossing the river on the blocks of ice."

"I have read it twice, your ladyship," was Jane's earnestly regretful response, "and most awful it is, and made me and mother cry beyond words. And I suppose it is the poor creature's colour that's against her, and I'm trying to be kind to her, but I must own that she makes me nervous. She asks me such a lot of questions in her queer way, and stares at me so quiet. She actually asked me quite sudden the other day if I loved the big Mem Sahib. I didn't know what she *could* mean at first, but after a while I found out it was her Indian way of meaning your ladyship, and she didn't intend disrespect, because she spoke of you most humble afterwards, and called his lordship the Heaven born."

"Be as kind as you can to her, Jane," instructed her mistress. "And take her a nice walk occasionally. I daresay she feels very homesick here."

What Ameerah said to her mistress was that these English servant women were pigs and devils, and could conceal nothing from those who chose to find out things from them. If Jane had known that the Ayah could have told her of every movement she made during the day or night, of her up-gettings and down-lyings, of the hour and moment of every service done for the big Mem Sahib, of why and how and when and where each thing was done, she would have been frightened indeed.

One day, it is true, she came into Lady Walderhurst's sleeping apartment to find Ameerah standing in the middle of it looking round its contents with restless, timid, bewildered eyes. She wore, indeed, the manner of an alarmed creature who did not know how she had got there.

"What are you doing here?" demanded Jane. "You have no right in this part of the house. You're taking a great liberty, and your mistress will be angry."

"My Mem Sahib asked for a book," the Ayah quite shivered in her alarmed confusion. "Your Mem Sahib said it was here. They did not order me, but I thought I would come to you. I did not know it was forbidden."

"What was the book?" inquired Jane severely. "I will take it to her ladyship."

But Ameerah was so frightened that she had forgotten the name, and when Jane knocked at the door of Mrs. Osborn's boudoir, it was empty, both the ladies having gone into the garden.

But Ameerah's story was quite true, Lady Walderhurst said in the evening when Jane spoke of the matter as she dressed her for dinner. They had been speaking of a book containing records of certain historical Walderhursts. It was one Emily had taken from the library to read in her bedroom.

"We did not ask her to go for it. In fact I did not know the woman was within hearing. She moves about so noiselessly one frequently does not know when she is near. Of course she meant very well, but she does not know our English ways."

"No, my lady, she does not," said Jane, respectfully but firmly. "I took the liberty of telling her she must keep to her own part of the house unless required by your ladyship."

"You mustn't frighten the poor creature," laughed her mistress. She was rather touched indeed by the slavish desire to please and do service swiftly which the Ayah's blunder seemed to indicate. She had wished to save her mistress even the trouble of giving the order. That was her Oriental way, Emily thought, and it was very affectionate and child-like.

Being reminded of the book again, she carried it down herself into the drawing-room. It was a volume she was fond of because it recorded romantic stories of certain noble dames of Walderhurst lineage.

Her special predilection was a Dame Ellena, who, being left with but few servitors in attendance during her lord's absence from his castle on a foraging journey into an enemy's country, had defended the stronghold boldly against the attack of a second enemy who had adroitly seized the opportunity to forage for himself. In the cellars had been hidden treasure recently acquired by the usual means, and knowing this, Dame Ellena had done splendid deeds, marshalling her small forces in such way as deceived the attacking party and showing herself in scorn upon the battlements, a fierce, beauteous woman about to give her lord an heir, yet fearing naught, and only made more fierce and full of courage by this fact. The son, born but three weeks later, had been the most splendid and savage fighter of his name, and a giant in build and strength.

"I suppose," Emily said when they discussed the legend after dinner, "I suppose she felt that she could do *anything*," with her italics. "I daresay *nothing* could make her afraid, but the thought that something might go wrong while her husband was away. And strength was given her."

She was so thrilled that she got up and walked across the room with quite a fine sweep of heroic movement in her momentary excitement. She held her head up and smiled with widening eyes.

But she saw Captain Osborn drag at his black moustache to hide an unattractive grin, and she was at once abashed into feeling silly and shy. She sat down again with awkward self-consciousness.

"I'm afraid I'm making you laugh at me," she apologised, "but that story always gives me such a romantic feeling. I like her so."

"Oh! not at all, not all," said Osborn. "I was not laughing really; oh no!"

But he had been, and had been secretly calling her a sentimental, ramping idiot.

It was a great day for Jane Cupp when her mother arrived at Palstrey Manor. It was a great day for Mrs. Cupp also. When she descended from the train at the little country station, warm and somewhat flushed by her emotions and the bugled splendours of her best bonnet and black silk mantle, the sight of Jane standing neatly upon the platform almost overcame her. Being led to his lordship's own private bus, and seeing her trunk surrounded by the attentions of an obsequious station-master and a liveried young man, she was conscious of concealing a flutter with dignified reserve.

"My word, Jane!" she exclaimed after they had taken their seats in the vehicle. "My word, you look as accustomed to it as if you had been born in the family."

But it was when, after she had been introduced to the society in the servants' hall, she was settled in her comfortable room next to Jane's own that she realised to the full that there were features of her position which marked it with importance almost startling. As Jane talked to her, the heat of the genteel bonnet and beaded mantle had nothing whatever to do with the warmth which moistened her brow.

"I thought I'd keep it till I saw you, mother," said the girl decorously. "I know what her ladyship feels about being talked over. If I was a lady myself, I shouldn't like it. And I know how deep you'll feel it, that when the doctor advised her to get an experienced married person to be at hand, she said in that dear way of hers, 'Jane, if your uncle could spare your mother, how I should like to have her. I've never forgot her kindness in Mortimer Street.'"

Mrs. Cupp fanned her face with a handkerchief of notable freshness.

"If she was Her Majesty," she said, "she couldn't be more sacred to me, nor me more happy to be allowed the privilege."

Jane had begun to put her mother's belongings away. She was folding and patting a skirt on the bed. She fussed about a little nervously and then lifted a rather embarrassed face.

"I'm glad you *are* here, mother," she said. "I'm thankful to *have* you!"

Mrs. Cupp ceased fanning and stared at her with a change of expression. She found herself involuntarily asking her next question in a half whisper.

"Why, Jane, what is it?"

Jane came nearer.

"I don't know," she answered, and her voice also was low. "Perhaps I'm silly and overanxious, because I *am* so fond of her. But that Ameerah, I actually dream about her."

"What! The black woman?",

"If I was to say a word, or if you did, and we was wrong, how should we feel? I've kept my nerves to myself till I've nearly screamed sometimes. And my lady would be so hurt if she knew. But—well," in a hurried outburst, "I do wish his lordship was here, and I do wish the Osborns wasn't. I do wish it, I tell you that."

"Good Lord!" cried Mrs. Cupp, and after staring with alarmed eyes a second or so, she wiped a slight dampness from her upper lip.

She was of the order of female likely to take a somewhat melodramatic view of any case offering her an opening in that direction.

"Jane!" she gasped faintly, "do you think they'd try to take her life?"

"Goodness, no!" ejaculated Jane, with even a trifle of impatience. "People like them daren't. But suppose they was to try to, well, to upset her in some way, what a thing for them it would be."

After which the two women talked together for some time in whispers, Jane bringing a chair to place opposite her mother's. They sat knee to knee, and now and then Jane shed a tear from pure nervousness. She was so appalled by the fear of making a mistake which, being revealed by some chance, would bring confusion upon and pain of mind to her lady.

"At all events," was Mrs. Cupp's weighty observation when their conference was at an end, "here we both are, and two pairs of eyes and ears and hands and legs is a fat lot better than one, where there's things to be looked out for."

Her training in the matter of subtlety had not been such as Ameerah's, and it may not be regarded as altogether improbable that her observation of the Ayah was at times not too adroitly concealed, but if the native woman knew that she was being remarked, she gave no sign of her knowledge. She performed her duties faithfully and silently, she gave no trouble, and showed a gentle subservience and humbleness towards the white servants which won immense approbation. Her manner towards Mrs. Cupp's self was marked indeed by something like a tinge of awed deference, which, it must be confessed, mollified the good woman, and awakened in her a desire to be just and lenient even to the dark of skin and alien of birth.

"She knows her betters when she sees them, and has pretty enough manners for a black," the object of her respectful obeisances remarked. "I wonder if she's ever heard of her Maker, and if a little brown Testament with good print wouldn't be a good thing to give her?"

This boon was, in fact, bestowed upon her as a gift. Mrs. Cupp bought it for a shilling at a small shop in the village. Ameerah, in whose dusky being was incorporated the occult faith of lost centuries, and whose gods had been gods through mystic ages, received the fat, little brown book with down-dropped lids and grateful obeisance. These were her words to her mistress:

"The fat old woman with protruding eyes bestowed it upon me. She says it is the book of her god. She has but one. She wishes me to worship him. Am I a babe to worship such a god as would please her. She is old, and has lost her mind."

Lady Walderhurst's health continued all that could be desired. She arose smiling in the morning, and bore her smile about with her all day. She walked much in the gardens, and spent long, happy hours sewing in her favourite sitting-room. Work which she might have paid other women to do, she did with her own hands for the mere sentimental bliss of it. Sometimes she sat with Hester and sewed, and Hester lay on a sofa and stared at her moving hands.

"You know how to do it, don't you?" she once said.

"I was obliged to sew for myself when I was so poor, and this is delightful," was Emily's answer.

"But you could buy it all and save yourself the trouble."

Emily stroked her bit of cambric and looked awkward.

"I'd rather not," she said.

Well as she was, she began to think she did not sleep quite so soundly as had been habitual with her. She started up in bed now and again as if she had been disturbed by some noise, but when she waited and listened she heard nothing. At least this happened on two or three occasions. And then one night, having been lying folded in profound, sweet sleep, she sprang up in the black darkness, wakened by an actual, physical reality of sensation, the soft laying of a hand upon her naked side,—that, and nothing else.

"What is that? Who is there?" she cried. "Someone is in the room!"

Yes, someone was there. A few feet from her bed she heard a sobbing sigh, then a rustle, then followed silence. She struck a match and, getting up, lighted candles. Her hand shook, but she remembered that she must be firm with herself.

"I must not be nervous," she said, and looked the room over from end to end.

But it contained no living creature, nor any sign that living creature had entered it since she had lain down to rest. Gradually the fast beating of her heart had slackened, and she passed her hand over her face in bewilderment.

"It wasn't like a dream at all," she murmured; "it really wasn't. I *felt* it."

Still as absolutely nothing was to be found, the sense of reality diminished somewhat, and being so healthy a creature, she regained her composure, and on going back to bed slept well until Jane brought her early tea.

Under the influence of fresh morning air and sunlight, of ordinary breakfast and breakfast talk with the Osborns, her first convictions receded so far that she laughed a little as she related the incident.

"I never had such a real dream in my life," she said; "but it must have been a dream."

"One's dreams are very real sometimes," said Hester.

"Perhaps it was the Palstrey ghost," Osborn laughed. "It came to you because you ignore it." He broke off with a slight sudden start and stared at her a second questioningly. "Did you say it put its hand on your side?" he asked.

"Don't tell her silly things that will frighten her. How ridiculous of you," exclaimed Hester sharply. "It's not proper."

Emily looked at both of them wonderingly.

"What do you mean?" she said. "I don't believe in ghosts. It won't frighten me, Hester. I never even heard of a Palstrey ghost."

"Then I am not going to tell you of one," said Captain Osborn a little brusquely, and he left his chair and went to the sideboard to cut cold beef.

He kept his back towards them, and his shoulders looked uncommunicative and slightly obstinate. Hester's face was sullen. Emily thought it sweet of her to care so much, and turned upon her with grateful eyes.

"I was only frightened for a few minutes, Hester," she said. "My dreams are not vivid at all, usually."

But howsoever bravely she ignored the shock she had received, it was not without its effect, which was that occasionally there drifted into her mind a recollection of the suggestion that Palstrey had a ghost. She had never heard of it, and was in fact of an orthodoxy so ingenuously entire as to make her feel that belief in the existence of such things was a sort of defiance of ecclesiastical laws. Still, such stories were often told in connection with old places, and it was natural to wonder what features marked this particular legend. Did it lay hands on people's sides when they were asleep? Captain Osborn had asked his question as if with a sudden sense of recognition. But she would not let herself think of the matter, and she would not make inquiries.

The result was that she did not sleep well for several nights. She was annoyed at herself, because she found that she kept lying awake as if listening or waiting. And it was not a good thing to lose one's sleep when one wanted particularly to keep strong.

Jane Cupp during this week was, to use her own words, "given quite a turn" by an incident which, though a small matter, might have proved untoward in its results.

The house at Palstrey, despite its age, was in a wonderful state of preservation, the carved oak balustrades of the stairways being considered particularly fine.

"What but Providence," said Jane piously, in speaking to her mother the next morning, "made me look down the staircase as I passed through the upper landing just before my lady was going down to dinner. What but Providence I couldn't say. It certainly wasn't because I've done it before that I remember. But just that one evening I was obliged to cross the landing for something, and my eye just lowered itself by accident, and there it was!"

"Just where it would have tripped her up. Good Lord! it makes my heart turn over to hear you tell it. How big a bit of carving was it?" Mrs. Cupp's opulent chest trimmings heaved.

"Only a small piece that had broken off from old age and worm-eatenness, I suppose, but it had dropped just where she wouldn't have caught sight of it, and ten to one would have stepped on it and turned her ankle and been thrown from the top to the bottom of the whole flight. Suppose I *hadn't* seen it in time to pick it up before she went down. Oh, dear! Oh, dear! Mother!"

"I should say so!" Mrs. Cupp's manner approached the devout. This incident it was which probably added to Jane's nervous sense of responsibility. She began to watch her mistress's movements with hyper-sensitive anxiety. She fell into the habit of going over her bedroom two or three times a day, giving a sort of examination to its contents.

"Perhaps I'm so fond of her that it's making me downright silly," she said to her mother; "but it seems as if I can't help it. I feel as if I'd like to know everything she does, and go over the ground to make sure of it before she goes anywhere. I'm so proud of her, mother; I'm just as proud as if I was some connection of the family, instead of just her maid. It'll be such a splendid thing if she keeps well and everything goes as it should. Even people like us can see what it means to a gentleman that can go back nine hundred years. If I was Lady Maria Bayne, I'd be here and never leave her. I tell you nothing could drive me from her."

"You are well taken care of," Hester had said. "That girl is devoted to you. In her lady's maid's way she'd fight for your life."

"I think she is as faithful to me as Ameerah is to you," Emily answered. "I feel sure Ameerah would fight for you."

Ameerah's devotion in these days took the form of a deep-seated hatred of the woman whom she regarded as her mistress's enemy.

"It is an evil thing that she should take this place," she said. "She is an old woman. What right hath she to think she may bear a son. Ill luck will come of it. She deserves any ill fortune which may befall her."

"Sometimes," Lady Walderhurst once said to Osborn, "I feel as if Ameerah disliked me. She looks at me in such a curious, stealthy way."

"She is admiring you," was his answer. "She thinks you are something a little supernatural, because you are so tall and have such a fresh colour."

There was in the park at Palstrey Manor a large ornamental pool of water, deep and dark and beautiful because of the age and hugeness of the trees which closed around it, and the water plants which encircled and floated upon it. White and yellow flags and brown velvet rushes grew thick about its edge, and water-lilies opened and shut upon its surface. An avenue of wonderful limes led down to a flight of mossy steps, by which in times gone by people had descended to the boat which rocked idly in the soft green gloom. There was an island on it, on which roses had been planted and left to run wild; early in the year daffodils and other spring flowers burst up through the grass and waved scented heads. Lady Walderhurst had discovered the place during her honeymoon, and had loved it fondly ever since. The avenue leading to it was her favourite walk; a certain seat under a tree on the island her favourite resting-place.

"It is so still there," she had said to the Osborns. "No one ever goes there but myself. When I have crossed the little old bridge and sit down among the greenness with my book or work, I feel as if there was no world at all. There is no sound but the rustle of the leaves and the splash of the moor-hens who come to swim about. They don't seem to be afraid of me, neither do the thrushes and robins. They know I shall only sit still and watch them. Sometimes they come quite near."

She used, in fact, to take her letter-writing and sewing to the sweet, secluded place and spend hours of pure, restful bliss. It seemed to her that her life became more lovely day by day.

Hester Osborn

Hester did not like the pool. She thought it too lonely and silent. She preferred her beflowered boudoir or the sunny garden. Sometimes in these days she feared to follow her own thoughts. She was being pushed—pushed towards the edge of her precipice, and it was only the working of Nature that she should lose her breath and snatch at strange things to stay herself. Between herself and her husband a sort of silence had grown. There were subjects of which they never spoke, and yet each knew that the other's mind was given up to thought of them day and night. There were black midnight hours when Hester, lying awake in her bed, knew that Alec lay awake in his also. She had heard him many a time turn over with a caught breath and a smothered curse. She did not ask herself what he was thinking of. She knew. She knew because she was thinking of the same things herself. Of big, fresh, kind Emily Walderhurst lost in her dreams of exultant happiness which never ceased to be amazed and grateful to prayerfulness; of the broad lands and great, comfortable houses; of all it implied to be the Marquis of Walderhurst or his son; of the long, sickening voyage back to India; of the hopeless muddle of life in an ill-kept bungalow; of wretched native servants, at once servile and stubborn and given to lies and thefts. More than once she was forced to turn on her face that she might smother her frenzied sobs in her pillow.

It was on such a night—she had awakened from her sleep to notice such stillness in Osborn's adjoining room, that she thought him profoundly asleep—that she arose from her bed to go and sit at her open window.

She had not been seated there many minutes before she became singularly conscious, she did not know how, of some presence near her among the bushes in the garden below. It had indeed scarcely seemed to be sound or movement which had attracted her attention, and yet it must have been one or both, for she involuntarily turned to a particular spot.

Yes, something, someone, was standing in a corner, hidden by shrubbery. It was the middle of the night, and people were meeting. She sat still and almost breathless. She could hear nothing and saw nothing but, between the leafage, a dim gleam of white. Only Ameerah wore white. After a few seconds' waiting she began to think a strange thing, though she presently realised that, taking all things into consideration, it was not strange at all. She got up very noiselessly and stole into her husband's room. He was not there; the bed was empty, though he had slept there earlier in the night.

She went back to her own bed and got into it again. In ten minutes' time Captain Osborn crept upstairs and returned to bed also. Hester made no sign and did not ask any questions. She knew he would have told her nothing, and also she did not wish to hear. She had seen him speaking to Ameerah in the lane a few days before, and now that he was meeting her in the night she knew that she need not ask herself what the subject of their consultation might be. But she looked haggard in the morning.

Lady Walderhurst herself did not look well, For the last two or three nights she had been starting from her sleep again with that eerie feeling of being wakened by someone at her bedside, though she had found no one when she had examined the room on getting up.

"I am sorry to say I am afraid I am getting a little nervous," she had said to Jane Cupp. "I will begin to take valerian, though it is really very nasty."

Jane herself had a somewhat harried expression of countenance. She did not mention to her mistress that for some days she had been faithfully following a line of conduct she had begun to mark out for herself. She had obtained a pair of list slippers and had been learning to go about softly. She had sat up late and risen from her bed early, though she had not been rewarded by any particularly marked discoveries. She had thought, however, that she observed that Ameerah did not look at her as much as had been her habit, and she imagined she rather avoided her. All she said to Lady Walderhurst was:

"Yes, my lady, mother thinks a great deal of valerian to quiet the nerves. Will you have a light left in your room to-night, my lady?"

"I am afraid I could not sleep with a light," her mistress answered. "I am not used to one."

She continued to sleep, disturbedly some nights, in the dark. She was not aware that on some of the nights Jane Cupp either slept or laid awake in the room nearest to her. Jane's own bedroom was in another part of the house, but in her quiet goings about in the list shoes she now and then saw things which made her nervously determined to be within immediate call.

"I don't say it isn't nerves, mother," she said, "and that I ain't silly to feel so suspicious of all sorts of little things, but there's nights when I couldn't stand it not to be quite near her."

CHAPTER SIXTEEN

The Lime Avenue was a dim, if lovely, place at twilight. When the sun was setting, broad lances of gold slanted through the branches and glorified the green spaces with mellow depths of light. But later, when the night was drawing in, the lines of grey tree-trunks, shadowed and canopied by boughs, suggested to the mind the pillars of some ruined cathedral, desolate and ghostly.

Jane Cupp, facing the gloom of it during her lady's dinner-hour, and glancing furtively from side to side as she went, would have been awed by the grey stillness, even if she had not been in a timorous mood to begin with. In the first place, the Lime Avenue, which was her ladyship's own special and favourite walk, was not the usual promenade of serving-maids. Even the gardeners seldom set foot in it unless to sweep away dead leaves and fallen wood. Jane herself had never been here before. This evening she had gone absolutely because she was following Ameerah.

She was following Ameerah because, during the afternoon tea-hour in the servants' hall, she had caught a sentence or so in the midst of a gossiping story, which had made her feel that she should be unhappy if she did not go down the walk and to the water-side,—see the water, the boat, the steps, everything.

"My word, mother!" she had said, "it's a queer business for a respectable girl that's maid in a great place to be feeling as if she had to watch black people, same as if she was in the police, and not daring to say a word; for if I did say a word, Captain Osborn's clever enough to have me sent away from here in a jiffy. And the worst of it is," twisting her hands together, "there *mayn't* be *anything* going on really. If they were as innocent as lambs they couldn't act any different; and just the same, things *might* have happened by accident."

"That's the worst of it," was Mrs. Cupp's fretted rejoinder. "Any old piece of carving might have dropped out of a balustrade, and any lady that wasn't well might have nightmare and be disturbed in her sleep."

"Yes," admitted Jane, anxiously, "that is the worst of it. Sometimes I feel so foolish I'm all upset with myself."

The gossip in servants' halls embraces many topics. In country houses there is naturally much to be said of village incidents, of the scandals of cottages and the tragedies of farms. This afternoon, at one end of the table the talk had been of a cottage scandal which had verged on tragedy. A handsome, bouncing, flaunting village girl had got into that "trouble" which had been anticipated for her by both friends and enemies for some time. Being the girl she was, much venomous village social stir had resulted. It had been predicted that she would "go up to London," or that she would drown herself, having an impudent high spirit which brought upon her much scornful and derisive flouting on her evil day. The manor servants knew a good deal of her, because she had been for a while a servant at The Kennel Farm, and had had a great fancy for Ameerah, whom it had pleased her to make friends with. When she fell suddenly ill, and for days lay at the point of death, there was a stealthy general opinion that Ameerah, with her love spells and potions, could have said much which might have been enlightening, if she had chosen. The girl had been in appalling danger. The village doctor, who had been hastily called in, had at one moment declared that life had left her body. It was, in fact, only Ameerah who had insisted that she was not dead. After a period of prostration, during which she seemed a corpse, she had slowly come back to earthly existence. The graphic descriptions of the scenes by her bedside, of her apparent death, her cold and bloodless body, her lagging and ghastly revival to consciousness, aroused in the servants' hall a fevered interest. Ameerah was asked questions, and gave such answers as satisfied herself if not her interlocutors. She was perfectly aware of the opinions of her fellow servitors. She knew all about them while they knew nothing whatsoever about her. Her limited English could be used as a means of baffling them. She smiled, and fell into Hindustani when she was pressed.

Jane Cupp heard both questions and answers. Ameerah professed to know nothing but such things as the whole village knew. Towards the end of the discussion, however, in a mixture of broken English and Hindustani, she conveyed that she had believed that the girl would drown herself. Asked why, she shook her head, then said that she had seen her by the Mem Sahib's lake at the end of the trees. She had asked if the water was deep enough, near the bridge, to drown. Ameerah had answered that she did not know.

There was a general exclamation. They all knew it was deep there. The women shuddered as they remembered how deep they had been told it was at that particular spot. It was said that there was no bottom to it. Everybody rather revelled in the gruesomeness of the idea of a bottomless piece of water. Someone remembered that there was a story about it. As much as ninety years ago two young labourers on the place had quarrelled about a young woman. One day, in the heat of jealous rage one had seized the other and literally thrown him into the pond. He had never been found. No drags could reach his body. He had sunk into the blackness for ever.

Ameerah sat at the table with downcast eyes. She had a habit of sitting silent with dropped eyes, which Jane could not bear. As she drank her tea she watched her in spite of herself.

After a few minutes had passed, her appetite for bread and butter deserted her. She got up and left the hall, looking pale.

The mental phases through which she went during the afternoon ended in her determination to go down the avenue and to the water's side this evening. It could be done while her ladyship and her guests were at dinner. This evening the Vicar and his wife and daughter were dining at the Manor.

Jane took in emotionally all the mysterious silence and dimness of the long tree-pillared aisle, and felt a tremor as she walked down it, trying to hold herself in hand by practical reflections half whispered.

"I'm just going to have a look, to make sure," she said, "silly or not. I've got upset through not being able to help watching that woman, and the way to steady my nerves is to make sure I'm just giving in to foolishness."

She walked as fast as she could towards the water. She could see its gleam in the dim light, but she must pass a certain tree before she could see the little bridge itself.

"My goodness! What's that?" she said suddenly. It was something white, which rose up as if from the ground, as if from the rushes growing at the water's edge.

Just a second Jane stood, and choked, and then suddenly darted forward, running as fast as she could. The white figure merely moved slowly away among the trees. It did not run or seem startled, and as Jane ran she caught it by its white drapery, and found herself, as she had known she would, dragging at the garments of Ameerah. But Ameerah only turned round and greeted her with a welcoming smile, mild enough to damp any excitement.

"What are you doing here?" Jane demanded. "Why do you come to this place?"

Ameerah answered her with simple fluency in Hindustani, with her manner of not realising that she was speaking to a foreigner who could not understand her. What she explained was that, having heard that Jane's Mem Sahib came here to meditate on account of the stillness, she herself had formed the habit of coming to indulge in prayer and meditation when the place was deserted for the day. She commended the place to Jane, and to Jane's mother, whom she believed to be holy persons given to devotional exercises. Jane shook her.

"I don't understand a word you say," she cried. "You know I don't. Speak in English."

Ameerah shook her head slowly, and smiled again with patience. She endeavoured to explain in English which Jane was sure was worse than she had ever heard her use before. Was it forbidden that a servant should come to the water?

She was far too much for Jane, who was so unnerved that she burst into tears.

"You are up to some wickedness," she sobbed; "I know you are. You're past bearing. I'm going to write to people that's got the right to do what I daren't. I'm going back to that bridge."

Ameerah looked at her with a puzzled amiability for a few seconds. She entered into further apologies and explanations in Hindustani. In the midst of them her narrow eyes faintly gleamed, and she raised a hand.

"They come to us. It is your Mem Sahib and her people. Hear them."

She spoke truly. Jane had miscalculated as to her hour, or the time spent at the dinner-table had been shorter than usual. In fact, Lady Walderhurst had brought her guests to see the young moon peer through the lime-trees, as she sometimes did when the evening was warm.

Jane Cupp fled precipitately. Ameerah disappeared also, but without precipitation or any sign of embarrassment.

* * * * *

"You look as if you had not slept well, Jane," Lady Walderhurst remarked in the morning as her hair was being brushed. She had glanced into the glass and saw that it reflected a pale face above her own, and that the pale face had red rims to its eyes.

"I have been a bit troubled by a headache, my lady," Jane answered.

"I have something like a headache myself." Lady Walderhurst's voice had not its usual cheerful ring. Her own eyes looked heavy. "I did not rest well. I have not rested well for a week. That habit of starting from my sleep feeling that some sound has disturbed me is growing on me. Last night I dreamed again that someone touched my side. I think I shall be obliged to send for Sir Samuel Brent."

"My lady," exclaimed Jane feverishly, "if you would—if you would."

Lady Walderhurst's look at her was nervous and disturbed.

"Do you—does your mother think I am not as well as I should be, Jane?" she said.

Jane's hands were actually trembling.

"Oh no, my lady. Oh no! But if Sir Samuel could be sent for, or Lady Maria Bayne, or—or his lordship—"

The disturbed expression of Lady Walderhurst's face changed to something verging on alarm. It was true that she began to be horribly frightened. She turned upon Jane, pallor creeping over her skin.

"Oh!" she cried, a sound of almost child-like fear and entreaty in her voice. "I am sure you think I am ill, I am sure you do. What—what is it?"

She leaned forward suddenly and rested her forehead on her hands, her elbows supported by the dressing-table. She was overcome by a shock of dread.

"Oh! if anything should go wrong!" in a faint half wail; "if anything *could* happen!" She could not bear the mere thought. It would break her heart. She had been so happy. God had been so good.

Jane was inwardly convulsed with contrition commingled with anger at her own blundering folly. Now it was she herself who had "upset" her ladyship, given her a fright that made her pale and trembling. What did she not deserve for being such a thoughtless fool. She might have known. She poured forth respectfully affectionate protestations.

"Indeed, I beg your pardon, my lady. Indeed, it's only my silliness! Mother was saying yesterday that she had never seen a lady so well and in as good spirits. I have no right to be here if I make such mistakes. Please, my lady—oh! might mother be allowed to step in a minute to speak to you?"

Emily's colour came, back gradually. When Jane went to her mother, Mrs. Cupp almost boxed her ears.

"That's just the way with girls," she said. "No more sense than a pack of cats. If you can't keep quiet you'd better just give up. Of course she'd think you meant they was to be sent for because we was certain she was a dying woman. Oh my! Jane Cupp, get away!"

She enjoyed her little interview with Lady Walderhurst greatly. A woman whose opinion was of value at such a time had the soundest reasons for enjoying herself. When she returned to her room, she sat and fanned herself with a pocket handkerchief and dealt judicially with Jane.

"What we've got to do," she said, "is to think, and think we will. Tell her things outright we must not, until we've got something sure and proved. Then we can call on them that's got the power in their hands. We can't call on them till we can show them a thing no one can't deny. As to that bridge, it's old enough to be easy managed, and look accidental if it broke. You say she ain't going there to-day. Well, this very night, as soon as it's dark enough, you and me will go down and have a look at it. And what's more, we'll take a man with us. Judd could be trusted. Worst comes to worst, we're only taking the liberty of making sure it's safe, because we know it *is* old and we're over careful."

As Jane had gathered from her, by careful and apparently incidental inquiry, Emily had had no intention of visiting her retreat. In the morning she had, in fact, not felt quite well enough. Her nightmare had shaken her far more on its second occurring. The stealthy hand had seemed not merely to touch, but to grip at her side, and she had been physically unable to rise for some minutes after her awakening. This experience had its physical and mental effects on her.

She did not see Hester until luncheon, and after luncheon she found her to be in one of her strange humours. She was often in these strange humours at this time. She wore a nervous and strained look, and frequently seemed to have been crying. She had new lines on her forehead between the eyebrows. Emily had tried in vain to rouse and cheer her with sympathetic feminine talk. There were days when she felt that for some reason Hester did not care to see her.

She felt it this afternoon, and not being herself at the high-water mark of cheerfulness, she was conscious of a certain degree of discouragement. She had liked her so much, she had wanted to be friends with her and to make her life an easier thing, and yet she appeared somehow to have failed. It was because she was so far from being a clever woman. Perhaps she might fail in other things because she was not clever. Perhaps she was never able to give to people what they wanted, what they needed. A brilliant woman had such power to gain and hold love.

After an hour or so spent in trying to raise the mental temperature of Mrs. Osborn's beflowered boudoir, she rose and picked up her little work-basket.

"Perhaps you would take a nap if I left you," she said. "I think I will stroll down to the lake."

She quietly stole away, leaving Hester on her cushions.

CHAPTER SEVENTEEN

A few minutes later a knock at the door being replied to by Hester's curt "Come in!" produced the modest entry of Jane Cupp, who had come to make a necessary inquiry of her mistress. "Her ladyship is not here; she has gone out." Jane made an altogether involuntary step forward. Her face became the colour of her clean white apron.

"Out!" she gasped.

Hester turned sharply round.

"To the lake," she said. "What do you mean by staring in that way?"

Jane did not tell her what she meant. She incontinently ran from the room without any shadow of a pretence at a lady's maid's decorum.

She fled through the rooms, to make a short cut to the door opening on to the gardens. Through that she darted, and flew across paths and flowerbeds towards the avenue of limes.

"She shan't get to the bridge before me," she panted. "She shan't, she shan't. I won't let her. Oh, if my breath will only hold out!"

She did not reflect that gardeners would naturally think she had gone mad. She thought of nothing whatever but the look in Ameerah's downcast eyes when the servants had talked of the bottomless water,—the eerie, satisfied, sly look. Of that, and of the rising of the white figure from the ground last night she thought, and she clutched her neat side as she ran.

The Lime Avenue seemed a mile long, and yet when she was running down it she saw Lady Walderhurst walking slowly under the trees carrying her touching little basket of sewing in her hand. She was close to the bridge.

"My lady! my lady!" she gasped out as soon as she dared. She could not run screaming all the way. "Oh, my lady! if you please!"

Emily heard her and turned round. Never had she been much more amazed in her life. Her maid, her well-bred Jane Cupp, who had not drawn an indecorous breath since assuming her duties, was running after her calling out to her, waving her hands, her face distorted, her voice hysteric.

Emily had been just on the point of stepping on to the bridge, her hand had been outstretched towards the rail. She drew back a step in alarm and stood staring. How strange everything seemed to-day. She began to feel choked and trembling.

A few seconds and Jane was upon her, clutching at her dress. She had so lost her breath that she was almost speechless.

"My lady," she panted. "Don't set foot on it; don't—don't, till we're sure."

"On—on what?"

Then Jane realised how mad she looked, how insane the whole scene was, and she gave way to her emotions. Partly through physical exhaustion and breathlessness, and partly through helpless terror, she fell on her knees.

"The bridge!" she said. "I don't care what happens to me so that no harm comes to you. There's things being plotted and planned that looks like accidents. The bridge would look like an accident if part of it broke. There's no bottom to the water. They were saying so yesterday, and *she* sat listening. I found her here last night."

"She! Her!" Emily felt as if she was passing through another nightmare.

"Ameerah," wailed poor Jane. "White ones have no chance against black. Oh, my lady!" her sense of the possibility that she might be making a fool of herself after all was nearly killing her. "I believe she would drive you to your death if she could do it, think what you will of me."

The little basket of needlework shook in Lady Walderhurst's hand. She swallowed hard, and without warning sat down on the roots of a fallen tree, her cheeks blanching slowly.

"Oh Jane!" she said in simple woe and bewilderment. "I don't understand any of it. How could—how *could* they want to hurt me!" Her innocence was so fatuous that she thought that because she had been kind to them they could not hate or wish to injure her.

But something for the first time made her begin to quail. She sat, and tried to recover herself. She put out a shaking hand to the basket of sewing. She could scarcely see it, because suddenly tears had filled her eyes.

"Bring one of the men here," she said, after a few moments. "Tell him that I am a little uncertain about the safety of the bridge."

She sat quite still while Jane was absent in search of the man. She held her basket on her knee, her hand resting on it. Her kindly, slow-working mind was wakening to strange thoughts. To her they seemed inhuman and uncanny. Was it because good, faithful, ignorant Jane had been rather nervous about Ameerah that she herself had of late got into a habit of feeling as if the Ayah was watching and following her. She had been startled more than once by finding her near when she had not been aware of her presence. She had, of course, heard Hester say that native servants often startled one by their silent, stealthy-seeming ways. But the woman's eyes had frightened her. And she had heard the story about the village girl.

She sat, and thought, and thought. Her eyes were fixed upon the moss-covered ground, and her breath came quickly and irregularly several times.

"I don't know what to do," she said. "I am sure—if it is true—I don't know what to do."

The under-gardener's heavy step and Jane's lighter one roused her. She lifted her eyes to watch the pair as they came. He was a big, young man with a simple rustic face and big shoulders and hands.

"The bridge is so slight and old," she said to him, "that it has just occurred to me that it might not be quite safe. Examine it carefully to make sure."

The young man touched his forehead and began to look the supports over. Jane watched him with bated breath when he rose to his feet.

"They're all right on this side, my lady," he said. "I shall have to get in the boat to make sure of them that rest on the island."

He stamped upon the end nearest and it remained firm.

"Look at the railing well," said Lady Walderhurst. "I often stand and lean on it and—and watch the sunset."

She faltered at this point, because she had suddenly remembered that this was a habit of hers, and that she had often spoken of it to the Osborns. There was a point on the bridge at which, through a gap in the trees, a beautiful sunset was always particularly beautiful. It was the right-hand rail facing these special trees she rested on when she watched the evening sky.

The big, young gardener looked at the left-hand rail and shook it with his strong hands.

"That's safe enough," he said to Jane.

"Try the other," said Jane.

He tried the other. Something had happened to it. It broke in his big grasp. His sunburnt skin changed colour by at least three shades.

"Lord A'mighty!" Jane heard him gasp under his breath. He touched his cap and looked blankly at Lady Walderhurst. Jane's heart seemed to herself to roll over. She scarcely dared look at her mistress, but when she took courage to do so, she found her so white that she hurried to her side.

"Thank you, Jane," she said rather faintly. "The sky is so lovely this afternoon that I meant to stop and look at it. I should have fallen into the water, which they say has no bottom. No one would have seen or heard me if you had not come."

She caught Jane's hand and held it hard. Her eyes wandered over the avenue of big trees, which no one but herself came near at this hour. It would have been so lonely, so lonely!

The gardener went away, still looking less ruddy than he had looked when he arrived on the spot. Lady Walderhurst rose from her seat on the mossy tree-trunk. She rose quite slowly.

"Don't speak to me yet, Jane," she said. And with Jane following her at a respectful distance, she returned to the house and went to her room to lie down.

There was nothing to prove that the whole thing was not mere chance, mere chance. It was this which turned her cold. It was all impossible. The little bridge had been entirely unused for so long a time, it had been so slight a structure from the first; it was old, and she remembered now that Walderhurst had once said that it must be examined and strengthened if it was to be used. She had leaned upon the rail often lately; one evening she had wondered if it seemed quite as steady as usual. What could she say, whom could she accuse, because a piece of rotten wood had given away.

She started on her pillow. It was a piece of rotten wood which had fallen from the balustrade upon the stairs, to be seen and picked up by Jane just before she would have passed down on her way to dinner. And yet, what would she appear to her husband, to Lady Maria, to anyone in the decorous world, if she told them that she believed that in a dignified English household, an English gentleman, even a deposed heir presumptive, was working out a subtle plot against her such as might adorn a melodrama? She held her head in her hands as her mind depicted to her Lord Walderhurst's countenance, Lady Maria's dubious, amused smile.

"She would think I was hysterical," she cried, under her breath. "He would think I was vulgar and stupid, that I was a fussy woman with foolish ideas, which made him ridiculous. Captain Os-born is of his family. I should be accusing him of being a criminal. And yet I might have been in the bottomless pond, in the bottomless pond, and no one would have known."

If it all had not seemed so incredible to her, if she could have felt certain herself, she would not have been overwhelmed with this sense of being baffled, bewildered, lost.

The Ayah who so loved Hester might hate her rival. A jealous native woman might be capable of playing stealthy tricks, which, to her strange mind, might seem to serve a proper end. Captain Osborn might not know. She breathed again as this thought came to her. He could not know; it would be too insane, too dangerous, too wicked.

And yet, if she had been flung headlong down the staircase, if the fall had killed her, where would have been the danger for the man who would only have deplored a fatal accident. If she had leaned upon the rail and fallen into the black depths of water below, what could have been blamed but a piece of rotten wood. She touched her forehead with her handkerchief because it felt cold and damp. There was no way out. Her teeth chattered.

"They may be as innocent as I am. And they may be murderers in their hearts. I can prove nothing, I can prevent nothing. Oh! *do* come home."

There was but one thought which remained clear in her mind. She must keep herself safe—she must keep herself safe. In the anguish of her trouble she confessed, by putting it into words, a thing which she had not confessed before, and even as she spoke she did not realise that her words contained confession.

"If I were to die now," she said with a touching gravity, "he would care very much."

A few moments later she said, "It does not matter what happens to me, how ridiculous or vulgar or foolish I seem, if I can keep myself safe—until after. I will write to him now and ask him to try to come back."

It was the letter she wrote after this decision which Osborn saw among others awaiting postal, and which he stopped to examine.

CHAPTER EIGHTEEN

Hester sat at the open window of her boudoir in the dark. She had herself put out the wax candles, because she wanted to feel herself surrounded by the soft blackness. She had sat through the dinner and heard her husband's anxious inquiries about the rotten handrail, and had watched his disturbed face and Emily's pale one. She herself had said but little, and had been glad when the time came that she could decently excuse herself and come away.

As she sat in the darkness and felt the night breath of the flowers in the garden, she was thinking of all the murderers she had ever heard of. She was reflecting that some of them had been quite respectable people, and that all of them must have lived through a period in which they gradually changed from respectable people to persons in whose brains a thought had worked which once they would have believed impossible to them, which they might have scouted the idea of their giving room to. She was sure the change must come about slowly. At first it would seem too mad and ridiculous, a sort of angry joke. Then the angry joke would return again and again, until at last they let it stay and did not laugh at it, but thought it over. Such things always happened because some one wanted, or did not want, something very much, something it drove them mad to think of being forced to live without, or with. Men who hated a woman and could not rid themselves of her, who hated the sight of her face, her eyes, her hair, the sound of her voice and step, and were rendered insane by her nearness and the thought that they never could be free from any of these things, had before now, commonplace or comparatively agreeable men, by degrees reached the point where a knife or a shot or a heavy blow seemed not only possible but inevitable. People who had been ill-treated, people who had faced horrors through want and desire, had reached the moment in which they took by force what Fate would not grant them. Her brain so whirled that she wondered if she was not a little delirious as she sat in the stillness thinking such strange things.

For weeks she had been living under a strain so intense that her feelings had seemed to cease to have any connection with what was normal.

She had known too much; and yet she had been certain of nothing at all.

But she and Alec were like the people who began with a bad joke, and then were driven and driven. It was impossible not to think of what might come, and of what might be lost for ever. If the rail had not been tried this afternoon, if big, foolish Emily Walderhurst had been lying peacefully among the weeds to-night!

"The end comes to everyone," she said. "It would have been all over in a few minutes. They say it isn't really painful."

Her lips quivered, and she pressed her hands tightly between her knees.

"That's a murderer's thought," she muttered querulously. "And yet I wasn't a bad girl to begin with."

She began to see things. The chief thing was a sort of vision of how Emily would have looked lying in the depths of the water among the weeds. Her brown hair would have broken loose, and perhaps tangled itself over her white face. Would her eyes be open and glazed, or half shut? And her childish smile, the smile that looked so odd on the face of a full-grown woman, would it have been fixed and seemed to confront the world of life with a meek question as to what she had done to people—why she had been drowned? Hester felt sure that was what her helpless stillness would have expressed.

How happy the woman had been! To see her go about with her unconsciously joyous eyes had sometimes been maddening. And yet, poor thing! why had she not the right to be happy? She was always trying to please people and help them. She was so good that she was almost silly. The day she had brought the little things from London to The Kennel Farm, Hester remembered that, despite her own morbid resentment, she had ended by kissing her with repentant tears. She heard again, in the midst of her delirious thoughts, the nice, prosaic emotion of her voice as she said:

"*Don't* thank me—don't. Just let us *enjoy* ourselves."

And she might have been lying among the long, thick weeds of the pond. And it would not have been the accident it would have appeared to be. Of that she felt sure. Brought face to face with this definiteness of situation, she began to shudder.

She went out into the night feeling that she wanted air. She was not strong enough to stand the realisation that she had become part of a web into which she had not meant to be knitted. No; she had had her passionate and desperate moments, but she had not meant things like this. She had almost hoped that disaster might befall, she had almost thought it possible that she would do nothing to prevent it—almost. But some things were too bad.

She felt small and young and hopelessly evil as she walked in the dark along a grass path to a seat under a tree. The very stillness of the night was a horror to her, especially when once an owl called, and again a dreaming bird cried in its nest.

She sat under the tree in the dark for at least an hour. The thick shadow of the drooping branches hid her in actual blackness and seclusion.

She said to herself later that some one of the occult powers she believed in had made her go out and sit in this particular spot, because there was a thing which was not to be, and she herself must come between.

When she at last rose it was with panting breath. She stole back to her room, and lighted with an unsteady hand a bedroom candle, whose flame flickered upon a distorted, little dark face. For as she had sat under the tree she had, after a while, heard whispering begin quite near her; had caught, even in the darkness, a gleam of white, and had therefore deliberately sat and listened.

*　*　*　*　*

There could be, to the purely normal geniality of Emily Walderhurst's nature, no greater relief than the recognition that a cloud had passed from the mood of another.

When Hester appeared the next morning at the breakfast-table, she had emerged from her humour of the day before and was almost affectionate in her amiability. The meal at an end, she walked with Emily in the garden.

She had never shown such interest in what pertained to her as she revealed this morning. Something she had always before lacked Emily recognised in her for the first time,—a desire to ask friendly questions, to verge on the confidential. They talked long and without reserve. And how pretty it was of the girl, Emily thought, to care so much about her health and her spirits, to be so interested in the details of her every-day life, even in the simple matter of the preparation and serving of her food, as if the merest trifle was of consequence. It had been unfair, too, to fancy that she felt no interest in Walderhurst's absence and return. She had noticed everything closely, and actually thought he ought to come back at once.

"Send for him," she said quite suddenly; "send for him now."

There was an eagerness expressed in the dark thinness of her face which moved Emily.

"It is dear of you to care so much, Hester," she said. "I didn't know you thought it mattered."

"He must come," said Hester. "That's all. Send for him."

"I wrote a letter yesterday," was Lady Walderhurst's meek rejoinder. "I got nervous."

"So did I get nervous," said Hester; "so did I."

That she was disturbed Emily could see. The little laugh she ended her words with had an excited ring in it.

During the Osborns' stay at Palstrey the two women had naturally seen a good deal of each other, but for the next two days they were scarcely separated at all. Emily, feeling merely cheered and supported by the fact that Hester made herself so excellent a companion, was not aware of two or three things. One was that Mrs. Osborn did not lose sight of her unless at such times as she was in the hands of Jane Cupp.

"I may as well make a clean breast of it," the young woman said. "I have a sense of responsibility about you that I haven't liked to speak of before. It's half hysterical, I suppose, but it has got the better of me."

"You feel responsible for *me*!" exclaimed Emily, with wondering eyes.

"Yes, I do," she almost snapped. "You represent so much. Walderhurst ought to be here. I'm not fit to take care of you."

"I ought to be taking care of you," said Emily, with gentle gravity. "I am the older and stronger. You are not nearly so well as I am."

Hester startled her by bursting into tears.

"Then do as I tell you," she said. "Don't go anywhere alone. Take Jane Cupp with you. You have nearly had two accidents. Make Jane sleep in your dressing-room."

Emily felt a dreary chill creep over her. That which she had felt in the air when she had slowly turned an amazed face upon Jane in the Lime Avenue, that sense of the strangeness of things again closed her in.

"I will do as you wish," she answered.

But before the next day closed all was made plain to her, all the awfulness, all the cruel, inhuman truth of things which seemed to lose their possibility in the exaggeration of proportion which made their incongruous ness almost grotesque.

The very prettiness of the flowered boudoir, the very softness of the peace in the velvet spread of garden before the windows, made it even more unreal.

That day, the second one, Emily had begun to note the new thing. Hester was watching her, Hester was keeping guard. And as she realised this, the sense of the abnormalness of things grew, and fear grew with it. She began to feel as if a wall were rising around her, built by unseen hands.

The afternoon, an afternoon of deeply golden sun, they had spent together. They had read and talked. Hester had said most. She had told stories of India,—curious, vivid, interesting stories, which seemed to excite her.

At the time when the sunlight took its deepest gold the tea-tray was brought in. Hester had left the room a short time before the footman appeared with it, carrying it with the air of disproportionate solemnity with which certain male domestics are able to surround the smallest service. The tea had been frequently served in Hester's boudoir of late. During the last week, however, Lady Walderhurst's share of the meal had been a glass of milk. She had chosen to take it because Mrs. Cupp had suggested that tea was "nervous." Emily sat down at the table and filled a cup for Hester. She knew she would return in a few moments, so set the cup before Mrs. Osborn's place and waited. She heard the young woman's footsteps outside, and as the door opened she lifted the glass of milk to her lips.

She was afterwards absolutely unable to describe to herself clearly what happened the next moment. In fact, it was the next moment that she saw Hester spring towards her, and the glass of milk had been knocked from her hand and rolled, emptying itself, upon the floor. Mrs. Osborn stood before her, clenching and unclenching her hands.

"Have you drunk any of it?" she demanded.

"No," Emily answered. "I have not."

Hester Osborn dropped into a chair and leaned forward, covering her face with her hands. She looked like a woman on the verge of an outbreak of hysteria, only to be held in check by a frenzied effort.

Lady Walderhurst, quite slowly, turned the colour of the milk itself. But she did nothing but sit still and gaze at Hester.

"Wait a minute." The girl was trying to recover her breath. "Wait till I can hold myself still. I am going to tell you now. I am going to tell you."

"Yes," Emily answered faintly.

It seemed to her that she waited twenty minutes before another word was spoken, that she sat quite that long looking at the thin hands which seemed to clutch the hidden face. This was a mistake arising from the intensity of the strain upon her nerves. It was scarcely five minutes before Mrs. Osborn lowered her hands and laid them, pressed tightly palm to palm, between her knees.

She spoke in a low voice, such a voice as a listener outside could not have heard.

"Do you know," she demanded, "what you represent to us—to me and to my husband—as you sit there?"

Emily shook her head. The movement of disclaimer was easier than speech. She felt a sort of exhaustion.

"I don't believe you do," said Hester. "You don't seem to realise anything. Perhaps it's because you are so innocent, perhaps it's because you are so foolish. You represent the thing that we have the right to *hate* most on earth."

"Do *you* hate me?" asked Emily, trying to adjust herself mentally to the mad extraordinariness of the situation, and at the same time scarcely understanding why she asked her question.

"Sometimes I do. When I do not I wonder at myself." The girl paused a second, looked down, as if questioningly, at the carpet, and then, lifting her eyes again, went on in a dragging, half bewildered voice: "When I do *not*, I actually believe it is because we are both—women together. Before, it was different."

The look which Walderhurst had compared to "that of some nice animal in the Zoo" came into Emily's eyes as two honest drops fell from them.

"Would *you* hurt me?" she faltered. "Could *you* let other people hurt me?"

Hester leaned further forward in her chair, widening upon her such hysterically insistent, terrible young eyes as made her shudder.

"Don't you *see*?" she cried. "*Can't* you see? But for *you* my son would be what Walderhurst is—my son, not yours."

"I understand," said Emily. "I understand."

"Listen!" Mrs. Osborn went on through her teeth. "Even for that, there are things I haven't the nerve to stand. I have thought I could stand them. But I can't. It does not matter why. I am going to tell you the truth. You represent too much. You have been too great a temptation. Nobody meant anything or planned anything at first. It all came by degrees. To see you smiling and enjoying everything and adoring that stilted prig of a Walderhurst put ideas into people's heads, and they grew because every chance fed them. If Walderhurst would come home—"

Lady Walderhurst put out her hand to a letter which lay on the table.

"I heard from him this morning," she said. "And he has been sent to the Hills because he has a little fever. He must be quiet. So you see he *cannot* come yet."

She was shivering, though she was determined to keep still.

"What was in the milk?" she asked.

"In the milk there was the Indian root Ameerah gave the village girl. Last night as I sat under a tree in the dark I heard it talked over. Only a few native women know it."

There was a singular gravity in the words poor Lady Walderhurst spoke in reply.

"That," she said, "would have been the cruelest thing of all."

Mrs. Osborn got up and came close to her.

"If you had gone out on Faustine," she said, "you would have met with *an accident*. It might or might not have killed you. But it would have been an accident. If you had gone downstairs before Jane Cupp saw the bit of broken balustrade you might have been killed—by accident again. If you had leaned upon the rail of the bridge you would have been drowned, and no human being could have been accused or blamed."

Emily gasped for breath, and lifted her head as if to raise it above the wall which was being slowly built round her.

"Nothing will be done which can be proved," said Hester Osborn. "I have lived among native people, and know. If Ameerah hated me and I could not get rid of her I should die, and it would all seem quite natural."

She bent down and picked up the empty glass from the carpet.

"It is a good thing it did not break," she said, as she put it on the tray. "Ameerah will think you drank the milk and that nothing will hurt you. You escape them always. She will be frightened."

As she said it she began to cry a little, like a child.

"Nothing will save *me*," she said. "I shall have to go back, I shall have to go back!"

"No, no!" cried Emily.

The girl swept away her tears with the back of a clenched hand.

"At first, when I hated you," she was even petulant and plaintively resentful, "I thought I could let it go on. I watched, and watched, and bore it. But the strain was too great. I broke down. I think I broke down one night, when something began to beat like a pulse against my side."

Emily got up and stood before her. She looked perhaps rather as she had looked when she rose and stood before the Marquis of Walderhurst on a memorable occasion, the afternoon on the moor. She felt almost quiet, and safe.

"What must I do?" she asked, as if she was speaking to a friend. "I am afraid. Tell me."

Little Mrs. Osborn stood still and stared at her. The most incongruous thought came to her mind. She found herself, at this weird moment, observing how well the woman held her stupid head, how finely it was set on her shoulders, and that in a modern

Royal Academy way she was rather like the Venus of Milo. It is quite out of place to think such things at such a time. But she found herself confronted with them.

"Go away," she answered. "It is all like a thing in a play, but I know what I am talking about. Say you are ordered abroad. Be cool and matter-of-fact. Simply go and hide yourself somewhere, and call your husband home as soon as he can travel."

Emily Walderhurst passed her hand over her forehead.

"It *is* like something in a play," she said, with a baffled, wondering face. "It isn't even respectable."

Hester began to laugh.

"No, it isn't even respectable," she cried. And her laughter was just in time. The door opened and Alec Osborn came in.

"What isn't respectable?" he asked.

"Something I have been telling Emily," she answered, laughing even a trifle wildly. "You are too young to hear such things. You must be kept respectable at any cost."

He grinned, but faintly scowled at the same time.

"You've upset something," he remarked, looking at the carpet.

"I have, indeed," said Hester. "A cup of tea which was half milk. It will leave a grease spot on the carpet. That won't be respectable."

When she had tumbled about among native servants as a child, she had learned to lie quickly, and she was very ready of resource.

CHAPTER NINETEEN

As she heard the brougham draw up in the wet street before the door, Mrs. Warren allowed her book to fall closed upon her lap, and her attractive face awakened to an expression of agreeable expectation, in itself denoting the existence of interesting and desirable qualities in the husband at the moment inserting his latch-key in the front door preparatory to mounting the stairs and joining her. The man who, after twenty-five years of marriage, can call, by his return to her side, this expression to the countenance of an intelligent woman is, without question or argument, an individual whose life and occupations are as interesting as his character and points of view.

Dr. Warren was of the mental build of the man whose life would be interesting and full of outlook if it were spent on a desert island or in the Bastille. He possessed the temperament which annexes incident and adventure, and the perceptiveness of imagination which turns a light upon the merest fragment of event. As a man whose days were filled with the work attendant upon the exercise of a profession from which can be withheld few secrets, and to which most mysteries explain themselves, his brain was the recording machine of impressions which might have stimulated to vividness of imagination a man duller than himself, and roused to feeling one of far less warm emotions.

He came into the room smiling. He was a man of fifty, of strong build, and masculine. He had good shoulders and good colour, and the eyes, nose, and chin of a man it would be a stupid thing to attempt to deal with in a blackguardly manner. He sat down in his chair by the fire and began to chat, as was his habit before he and his wife parted to dress for dinner. When he was out during the day he often looked forward to these chats, and made notes of things he would like to tell his Mary. During her day, which was given to feminine duties and pleasures, she frequently did the same thing. Between seven and eight in the evening they had delightful conversational opportunities. He picked up her book and glanced it over, he asked her a few questions and answered a few; but she saw it was with a somewhat preoccupied manner. She knew a certain remote look in his eye, and she waited to see him get up from his chair and begin to walk to and fro, with his hands in his pockets and his head thrown back. When, after having done this, he began in addition to whistle softly and draw his eyebrows together, she broke in upon him in the manner of merely following an established custom.

"I am perfectly sure," was her remark, "that you have come upon one of the Extraordinary Cases."

The last two words were spoken as with inverted commas. Of many deep interests he added to her existence, the Extraordinary Cases were among the most absorbing. He had begun to discuss them with her during the first year of their married life. Accident had thrown one of them into her immediate personal experience, and her clear-headed comprehension and sympathy in summing up singular evidence had been of such value to him that he had turned to her in the occurrence of others for the aid straightforward, mutual logic could give. She had learned to await the Extraordinary Case with something like eagerness. Sometimes, it was true, its incidents were painful; but invariably they were absorbing in their interest, and occasionally illuminating beyond description. Of names and persons it was not necessary she should hear anything— the drama, the ethics, were enough. With an absolute respect for his professional reserves, she asked no questions he could not reply to freely, and avoided even the innocent following of clues. The Extraordinary Case was always quite enough as it stood. When she saw the remotely speculative look in his eye, she suspected one, when he left his chair and paced the floor with that little air of restlessness, and ended with unconscious whistling which was scarcely louder than a breath, she felt that evidence enough had accumulated for her.

He stopped and turned round.

"My good Mary," he owned at once, "its extraordinariness consists in its baffling me by being so perfectly ordinary."

"Well, at least that is not frequent. What is its nature? Is it awful? Is it sad? Is it eccentric? Is it mad or sane, criminal or domestic?"

"It is nothing but suggestive, and that it suggests mystery to me makes me feel as if I myself, instead of a serious practitioner, am a professional detective."

"Is it a case in which you might need help?"

"It is a case in which I am impelled to give help, if it proves that it is necessary. She is such an exceedingly nice woman."

"Good, bad, or indifferent?"

"Of a goodness, I should say—of a goodness which might prevent the brain acting in the manner in which a brutal world requires at present that the human brain should act in self-defence. Of a goodness which may possibly have betrayed her into the most pathetic trouble."

"Of the kind—?" was Mrs. Warren's suggestion.

"Of that kind," with a troubled look; "but she is a married woman."

"She *says* she is a married woman."

"No. She does not say so, but she looks it. That's the chief feature of the case. Any woman bearing more obviously the stamp of respectable British matrimony than this one does, it has not fallen to me to look upon."

Mrs. Warren's expression was *intriguée* in the extreme. There was a freshness in this, at least.

"But if she bears the stamp as well as the name—! Do tell me all it is possible to tell. Come and sit down, Harold."

He sat down and entered into details.

"I was called to a lady who, though not ill, seemed fatigued from a hurried journey and, as it seemed to me, the effects of anxiety and repressed excitement. I found her in a third-class lodging-house in a third-class street. It was a house which had the air of a place hastily made inhabitable for some special reason. There were evidences that money had been spent, but that there had been no time to arrange things. I have seen something of the kind before, and when I was handed into my patient's sitting-room, thought I knew the type I should find. It is always more or less the same,—a girl or a very young woman, pretty and refined and frightened, or pretty and vulgar and 'carrying it off' with transparent pretences and airs and graces. Anything more remote from what I expected you absolutely cannot conceive."

"Not young and pretty?"

"About thirty-five or six. A fresh, finely built woman with eyes as candid as a six-year-old girl's. Quite unexplanatory and with the best possible manner, only sweetly anxious about her health. Her confidence in my advice and the earnestness of her desire to obey my least instructions were moving. Ten minutes' conversation with her revealed to me depths of long-secreted romance in my nature. I mentally began to swear fealty to her."

"Did she tell you that her husband was away?"

"What specially struck me was that it did not occur to her that her husband required stating, which was ingenuously impressive. She did not explain her mother or her uncles, why her husband? Her mental attitude had a translucent clearness. She wanted a medical man to take charge of her, and if she had been an amiable, un-brilliant lady who was a member of the royal house, she would have conversed with me exactly as she did."

"She was so respectable?"

"She was even a little Mid-Victorian, dear Mary; a sort of clean, healthy, Mid-Victorian angel."

"There's an incongruousness in the figure in connection with being obviously in hiding in a lodging-house street." And Mrs. Warren gave herself to reflection.

"I cannot make it as incongruous as she was. I have not told you all. I have saved to the last the feature which marked her most definitely as an Extraordinary Case. I suppose one does that sort of thing from a sense of drama."

"What else?" inquired Mrs. Warren, roused from her speculation.

"What respectable conclusion *could* one deduce from the fact that a letter lay on the table near her, sealed with an imposing coat of arms. One's eye having accidentally fallen on it, one could, of course, only avoid glancing at it again, so I recognised nothing definite. Also, when I was announced unexpectedly, I saw her quickly withdraw her hand from her lips. She had been kissing a ring she wore. I could not help seeing that afterwards. My good Mary, it was a ruby, of a size and colour which recalled the Arabian Nights."

Mrs. Warren began to resign herself.

"No," she said, "there is no respectable conclusion to be drawn. It is tragic, but prosaic. She has been governess or companion in some great house. She may be a well-born woman. It is ten times more hideous for her than if she were a girl. She has to writhe under knowing that both her friends and her enemies are saying that she had not the excuse of not having been old enough to know better."

"That might all be true," he admitted promptly. "It *would* be true if—but she is not writhing. She is no more unhappy than you or I. She is only anxious, and I could swear that she is only anxious about one thing. The moment in which I swore fealty to her was when she said to me, 'I want to be quite safe—until after. I do not care for myself. I will bear anything or do anything. Only one thing matters. I shall be such a good patient.' Then her eyes grew moist, and she closed her lips decorously to keep them from trembling.

"They're not usually like that," Mrs. Warren remarked.

"I have not found them so," he replied.

"Perhaps she believes the man will marry her."

There was odd unexpectedness in the manner in which Dr. Warren suddenly began to laugh.

"My dear wife, if you could see her! It is the incongruity of what we are saying which makes me laugh. With her ruby and her coronets and her lodging-house street, she is of an impeccableness! She does not even know she could be doubted. Fifteen years of matrimony spent in South Kensington, three girls in the schoolroom and four boys at Eton, could not have crystallised a more unquestionable serenity. And you are saying gravely, 'Perhaps she believes the man will marry her.' Whatsoever the situation is, I am absolutely sure that she has never asked herself whether he would or not."

"Then," Mrs. Warren answered, "it is the most Extraordinary Case we have had yet."

"But I have sworn fealty to her," was Warren's conclusion. "And she will tell me more later." He shook his head with an air of certainty. "Yes, she will feel it necessary to tell me later."

They went upstairs to dress for dinner, and during the remainder of the evening which they spent alone they talked almost entirely of the matter.

CHAPTER TWENTY

Lady Walderhurst's departure from Palstrey, though unexpected, had been calm and matter-of-fact. All the Osborns knew was that she had been obliged to go up to London for a day or two, and that when there, her physician had advised certain German baths. Her letter of explanation and apology was very nice. She could not return to the country before beginning her journey. It seemed probable that she would return with her husband, who might arrive in England during the next two months.

"Has she heard that he is coming back?" Captain Osborn asked his wife.

"She has written to ask him to come."

Osborn grinned.

"He will be obliged to her. He is tremendously pleased with his importance at this particular time, and he is just the sort of man—as we both know—to be delighted at being called back to preside over an affair which is usually a matter for old women."

But the letter he had examined, as it lay with the rest awaiting postal, he had taken charge of himself. He knew that one, at least, would not reach Lord Walderhurst. Having heard in time of the broken bridge-rail, he had been astute enough to guess that the letter written immediately after the incident might convey such impressions as might lead even his lordship to feel that it would be well for him to be at home. The woman had been frightened, and would be sure to lose her head and play the fool. In a few days she would calm down and the affair would assume smaller proportions. At any rate, he had chosen to take charge of this particular letter.

What he did not know, however, was that chance had played into his hands in the matter of temporarily upsetting Lord Walderhurst's rather unreliable digestion, and in altering his plans, by a smart, though not dangerous, attack of fever which had ended in his being ordered to a part of the hill country not faithfully reached by letters; as a result of which several communications from his wife went astray and were unduly delayed. At the time Captain Osborn was discussing him with Hester, he was taking annoyed care of himself with the aid of a doctor, irritated by the untoward disturbance of his arrangements, and giving, it is true, comparatively little thought to his wife, who, being comfortably installed at Palstrey Manor, was doubtless enjoyably absorbed in little Mrs. Osborn.

"What German baths does she intend going to?" Alec Osborn inquired.

Hester consulted the letter with a manner denoting but languid interest.

"It's rather like her that she doesn't go to the length of explaining," was her reply. "She has a way of telling you a great many things you don't care to know, and forgetting to mention those you are interested in. She is very detailed about her health, and her affection and mine. She evidently expects us to go back to The Kennel Farm, and deplores her inhospitality, with adjectives."

She did not look as if she was playing a part; but she was playing one, and doing it well. Her little way was that of a nasty-tempered, self-centred woman, made spiteful by being called upon to leave a place which suited her.

"You are not really any fonder of her than I am," commented Osborn, after regarding her speculatively a few moments. If he had been as sure of her as he had been of Ameerah—!

"I don't know of any reason for my being particularly fond of her," she said. "It's easy enough for a rich woman to be good-natured. It doesn't cost her enough to constitute a claim."

Osborn helped himself to a stiff whiskey and soda. They went back to The Kennel Farm the next day, and though it was his habit to consume a large number of "pegs" daily, the habit increased until there were not many hours in the day when he was normally sure of what he was doing.

The German baths to which Lady Walderhurst had gone were nearer to Palstrey than any one knew. They were only at a few hours' distance by rail.

When, after a day spent in a quiet London lodging, Mrs. Cupp returned to her mistress with the information that she had been to the house in Mortimer Street and found that the widow who had bought the lease and furniture was worn out with ill-luck and the uncertainty of lodgers, and only longed for release which was not ruin, Emily cried a little for joy.

"Oh, how I should like to be there!" she said. "It was such a dear house. No one would ever dream of my being in it. And I need have no one but you and Jane. I should be so safe and quiet. Tell her you have a friend who will take it, as it is, for a year, and pay her anything."

"I won't tell her quite that, my lady," Mrs. Cupp made sagacious answer. "I'll make her an offer in ready money down, and no questions asked by either of us. People in her position sometimes gets a sudden let that pays them better than lodgers. All classes has their troubles, and sometimes a decent house is wanted for a few months, where money can be paid. I'll make her an offer."

The outcome of which was that the widowed householder walked out of her domicile the next morning with a heavier purse and a lighter mind than she had known for many months. The same night, ingenuously oblivious of having been called upon to fill the role of a lady in genteel "trouble," good and decorous Emily Walderhurst arrived under the cover of discreet darkness in a cab, and when she found herself in the "best bedroom," which had once been so far beyond her means, she cried a little for joy again, because the four dull walls, the mahogany dressing-table, and ugly frilled pincushions looked so unmelodramatically normal and safe.

"It seems so home-like," she said; adding courageously, "it is a very comfortable place, really."

"We can make it much more cheerful, my lady," Jane said, with grateful appreciation. "And the relief makes it like Paradise." She was leaving the room and stopped at the door. "There's not a person, black or white, can get across the door-mat, past mother and me, until his lordship comes," she allowed herself the privilege of adding.

Emily felt a little nervous when she pictured to herself Lord Walderhurst crossing the door-mat of a house in Mortimer Street in search of his Marchioness. She had not yet had time to tell him the story of the episode of the glass of milk and Hester Osborn's sudden outburst. Every moment had been given to carefully managed arrangement for the journey which was to seem so natural. Hester's cleverness had suggested every step and had supported her throughout. But for Hester she was afraid she might have betrayed herself. There had been no time for writing. But when James received her letter (of late she had more than once thought of him as "James"), he would know the one thing that was important. And she had asked him to come to her. She had apologised for suggesting any alteration of his plans, but she had really asked him to come to her.

"I think he will come," she said to herself. "I do think he will. I shall be so glad. Perhaps I have not been sensible, perhaps I have not done the best thing, but if I keep myself safe until he comes back, that really seems what is most important."

Two or three days in the familiar rooms, attended only by the two friendly creatures she knew so well, seemed to restore the balance of life for her. Existence became comfortable and prosaic again. The best bedroom and the room in which she spent her days were made quite cheerful through Jane's enterprise and memories of the appointments of Palstrey. Jane brought her tea in the morning, Mrs. Cupp presided over the kitchen. The agreeable doctor, whose reputation they had heard so much of, came and went, leaving his patient feeling that she might establish a friendship. He looked so clever and so kind.

She began to smile her childlike smile again. Mrs. Cupp and Jane told each other in private that if she had not been a married lady, they would have felt that she was Miss Fox-Seton again. She looked so like herself, with her fresh colour and her nice, cheerful eyes. And yet to think of the changes there had been, and what they had gone through!

People in London know nothing—or everything—of their neighbours. The people who lived in Mortimer Street were of the hard-worked lodging-house keeping class, and had too many anxieties connected with butcher's bills, rent, and taxes, to be able to give much time to their neighbours. The life in the house which had changed hands had nothing noticeable about it. It looked from the outside as it had always looked. The door-steps were kept clean, milk was taken in twice a day, and local tradesmen's carts left things in the ordinary manner. A doctor occasionally called to see someone, and the only person who had inquired about the patient (she was a friendly creature, who met Mrs. Cupp at the grocer's, and exchanged a few neighbourly words) was told that ladies who lived in furnished apartments, and had nothing to do, seemed to find an interest in seeing a doctor about things working-women had no time to bother about. Mrs. Cupp's view seemed to be that doctor's visits and medicine bottles furnished entertainment. Mrs. Jameson had "as good a colour and as good an appetite as you or me," but she was one who "thought she caught cold easy," and she was "afraid of fresh air."

Dr. Warren's interest in the Extraordinary Case increased at each visit he made. He did not see the ruby ring again. When he had left the house after his first call, Mrs. Cupp had called Lady Walderhurst's attention to the fact that the ring was on her hand, and could not be considered compatible with even a first floor front in Mortimer Street. Emily had been frightened and had removed it.

"But the thing that upsets me when I hand him in," Jane said to her mother anxiously in private, "is the way she can't help looking. You know what I mean, mother,—her nice, free, *good* look. And we never *could* talk to her about it. We should have to let her know that it's more than likely he thinks she's just what she isn't. It makes me mad to think of it. But as it had to be, if she only looked a little awkward, or not such a lady, or a bit uppish and fretful, she would seem so much more real. And then there's another thing. You know she always *did* carry her head well, even when she was nothing but poor Miss Fox-Seton tramping about shopping with muddy feet. And now, having been a marchioness till she's got used to it, and knowing that she is one, gives her an innocent, stately look sometimes. It's a thing she doesn't know of herself, but I do declare that sometimes as she's sat there talking just as sweet as could be, I've felt as if I ought to say, 'Oh! if you please, my lady, if you *could* look not quite so much as if you'd got on a tiara.'"

"Ah!" and Mrs. Cupp shook her head, "but that's what her Maker did for her. She was born just what she looks, and she looks just what she was born,—a respectable female."

Whereby Dr. Warren continued to feel himself baffled.

"She only goes out for exercise after dark, Mary," he said. "Also in the course of conversation I have discovered that she believes every word of the Bible literally, and would be alarmed if one could not accept the Athanasian Creed. She is rather wounded and puzzled by the curses it contains, but she feels sure that it would be wrong to question anything in the Church Service. Her extraordinariness is wholly her incompatibleness."

Gradually they had established the friendship Emily had thought possible. Once or twice Dr. Warren took tea with her. Her unabashed and accustomed readiness of hospitality was as incompatible with her circumstances as all the rest. She had the ease of a woman who had amiably poured out tea for afternoon callers all her life. Women who were uncertain of themselves were not amiably at ease with small social amenities. Her ingenuous talk and her fervent italics were an absolute delight to the man who was studying her. He, too, had noticed the carriage of her head Jane Cupp had deplored.

"I should say she was well born," he commented to his wife. "She holds herself as no common woman could."

"Ah! I haven't a doubt that she is well born, poor soul."

"No, not 'poor soul.' No woman who is as happy as she is needs pity. Since she has had time to rest, she looks radiant."

In course of time, however, she was less radiant. Most people know something of waiting for answers to letters written to foreign lands. It seems impossible to calculate correctly as to what length of time must elapse before the reply to the letter one sent by the last mail can reach one. He who waits is always premature in the calculation he makes. The mail should be due at a certain date, one is so sure. The letter could be written on such a day and posted at once. But the date calculated for arrives, passes,—the answer has not come. Who does not remember?

Emily Walderhurst had passed through the experience and knew it well. But previously the letters she had sent had been of less vital importance. When the replies to them had lingered on their way she had, it is true, watched eagerly ', for the postman, and had lived restlessly between the arrivals of the mails, but she had taught herself resignation to the inevitable. Now life had altered its aspect and its significance. She had tried, with the aid of an untried imagination, to paint to herself the moments in which her husband would read the letter which told him what she had told. She had wondered if he would start, if he would look amazed, if his grey-brown eyes would light with pleasure! Might he not want to see her? Might he not perhaps write at once? She never could advance farther in her imagined reading of this reply than the first lines:

"MY DEAR EMILY,—The unexpected good news your letter contains has given me the greatest satisfaction. You do not perhaps know how strong my desire has been—"

She used to sit and flush with happiness when she reached this point. She so wished that she was capable of depicting to herself what the rest would be.

She calculated with the utmost care the probable date of the epistle's arrival. She thought she made sure of allowing plenty of time for all possible delays. The safety of her letters she had managed, with Hester's aid, to arrange for. They were forwarded to her bankers and called for. Only the letters from India were of any importance, and they were not frequent. She told herself that she must be even more than usually patient this time. When the letter arrived, if he told her he felt it proper that he should return, no part of the strange experience she had passed through would be of moment. When she saw his decorous, well-bred face and heard his correctly modulated voice, all else would seem like an unnatural dream.

In her relief at the decent composure of the first floor front in Mortimer Street the days did not seem at first to pass slowly. But as the date she had counted on drew near she could not restrain a natural restlessness. She looked at the clock and walked up and down the room a good deal. She was also very glad when night came and she could go to bed. Then she was glad when the morning arrived, because she was a day nearer to the end.

On a certain evening Dr. Warren said to his wife, "She is not so well to-day. When I called I found her looking pale and anxious. When I commented on the fact and asked how she was, she said that she had had a disappointment. She had been expecting an important letter by a mail arriving yesterday, and it had not come. She was evidently in low spirits."

"Perhaps she has kept up her spirits before because she believed the letter would come," Mrs. Warren speculated.

"She has certainly believed it would come."

"Do *you* think it will, Harold?"

"She thinks it will yet. She was pathetically anxious not to be impatient. She said she knew there were so many reasons for delay when people were in foreign countries and very much occupied."

"There are many reasons, I daresay," said Mrs. Warren with a touch of bitterness," but they are not usually the ones given to waiting, desperate women."

Dr. Warren stood upon the hearthrug and gazed into the fire, knitting his brows.

"She wanted to tell or ask me something this afternoon," he said, "but she was afraid. She looked like a good child in great trouble. I think she will speak before long."

She looked more and more like a good child in trouble as time passed. Mail after mail came in, and she received no letter. She did not understand, and her fresh colour died away. She spent her time now in inventing reasons for the non-arrival of her letter. None of them comprised explanations which could be disparaging in any sense to Walderhurst. Chiefly she clung to the fact that he had not been well. Anything could be considered a reason for neglecting letter writing if a man was not well. If his illness had become serious she would, of course, have heard from his doctor. She would not allow herself to contemplate that. But if he was languid and feverish, he might so easily put off writing from day to day. This was all the more plausible as a reason, since he had not been a profuse correspondent. He had only written when he had found he had leisure, with decent irregularity, so to speak.

At last, however, on a day when she had felt the strain of waiting greater than she had courage for, and had counted every moment of the hour which must elapse before Jane could return from her mission of inquiry, as she rested on the sofa she heard the girl mount the stairs with a step whose hastened lightness wakened in her an excited hopefulness.

She sat up with brightened face and eager eyes. How foolish she had been to fret. Now—now everything would be different. Ah! how thankful she was to God for being so good to her!

"I think you must have a letter, Jane," she said the moment the door opened. "I felt it when I heard your footstep."

Jane was touching in her glow of relief and affection.

"Yes, my lady, I have, indeed. And they said at the bank that it had come by a steamer that was delayed by bad weather."

Emily took the letter. Her hand shook, but it was with pleasure. She forgot Jane, and actually kissed the envelope before she opened it. It looked like a beautiful, long letter. It was quite thick.

But when she had opened it, she saw that the letter itself was not very long. Several extra sheets of notes or instructions, it did not matter what, seemed to be enclosed. Her hand shook so that she let them fall on the floor. She looked so agitated that Jane was afraid to do more than retire discreetly and stand outside the door.

In a few minutes she congratulated herself on the wisdom of not having gone downstairs. She heard a troubled exclamation of wonder, and then a call for herself.

"Jane, please, Jane!"

Lady Walderhurst was still sitting upon the sofa, but she looked pale and unsteady. The letter was in her hand, which rested weakly in her lap. It seemed as if she was so bewildered that she felt helpless.

She spoke in a tired voice.

"Jane," she said, "I think you will have to get me a glass of wine. I don't think I am going to faint, but I do feel so—so upset."

Jane was at her side kneeling by her.

"Please, my lady, lie down," she begged. "Please do."

But she did not lie down. She sat trembling and looking at the girl in a pathetic, puzzled fashion.

"I don't think," she quavered, "that his lordship can have received my letter. He can't have received it. He doesn't say anything. He doesn't say one word—"

She had been too healthy a woman to be subject to attacks of nerves. She had never fainted before in her life, and as she spoke she did not at all understand why Jane seemed to move up and down, and darkness came on suddenly in the middle of the morning.

Jane managed by main strength to keep her from falling from the sofa, and thanked Providence for the power vouchsafed to her. She reached the bell and rang it violently, and hearing it, Mrs. Cupp came upstairs with heavy swiftness.

CHAPTER TWENTY ONE

Naturally a perceptive and closely reasoning woman, Mrs. Warren's close intellectual intimacy with her husband had, in giving her the benefit of intercourse with a wide experience, added greatly to her power of reasoning by deduction. Warren frequently felt that his talk with her was something like consultation with a specially clever and sympathetic professional confrère. Her suggestions or conclusions were invariably worth consideration. More than once his reflection upon them had led him to excellent results. She made one night a suggestion with regard to the Extraordinary Case which struck him as being more than usually astute.

"Is she an intellectual woman?" she inquired.

"Not in the least. An unsparingly brilliant person might feel himself entitled to the right to call her stupid."

"Is she talkative?"

"Far from it. One of her charms is the nice respect she seems to feel for the remarks of others."

"And she is not excitable?"

"Rather the reverse. If excitability is liveliness, she is dull."

"I see," slowly, "you have not yet thought it possible that she might—well—be under some delusion."

Warren turned quickly and looked at her.

"It is wonderfully brilliant of you to have thought of it. A delusion?" He stood and thought it over.

"Do you remember," his wife assisted him with, "the complications which arose from young Mrs. Jerrold's running away, under similar circumstances, to Scotland and hiding herself in a shepherd's cottage under the impression that her husband was shadowing her with detectives? You recollect what a lovable woman she was, and what horror she felt of the poor fellow."

"Yes, yes. That was an Extraordinary Case too."

Mrs. Warren warmed with her subject.

"Here is a woman obviously concealing herself from the world in a lodging-house, plainly possessing money, owning a huge ruby ring, receiving documents stamped with imposing seals, taking exercise only by night, heart-wrung over the non-arrival of letters which are due. Every detail points to one painful, dubious situation. On the other hand, she presents to you the manner and aspect of a woman who is absolutely not dubious, and who is merely anxious on the one point a dubious person would be indifferent to. Isn't it, then, possible that over-wrought physical condition may have driven her to the belief that she is hiding from danger."

Dr. Warren was evidently following the thought seriously.

"She said," reflecting, "that all that mattered was that she should be safe. 'I want to keep safe.' That was it. You are very enlightening, Mary, always. I will go and see her again to-morrow. But," as the result of another memory, "how sane she seems!"

* * * * *

He was thinking of this possible aspect of the matter as he mounted the staircase of the house in Mortimer Street the next day. The stairway was of the ordinary lodging-house type, its dinginess somewhat alleviated by the fact that the Cupps had covered the worn carpet with clean warm-coloured felting. The yellowish marbled paper on the walls depressed the mind as one passed it; the indeterminate dun paint had defied fog for years. The whole house presented only such features as would encourage its proprietors to trust to the sufficing of infrequent re-decoration.

Jane had, however, made efforts in behalf of the drawing-room, in which her mistress spent her days. She had introduced palliations by degrees and with an unobtrusiveness which was not likely to attract the attention of neighbours unaccustomed to lavish delivery by means of furniture vans. She had brought in a rug or so, and had gradually replaced objects with such as were more pleasant to live with and more comfortable to use. Dr. Warren had seen the change wrought, and had noted evidences that money was not unobtainable. The maid also was a young woman whose manner towards her mistress was not merely respectful and well-bred, but suggestive of watchful affection bordering on reverence. Jane Cupp herself was a certificate of decorum and good standing. It was not such young women who secluded themselves with questionable situations. As she laid her hand on the drawing-room door to open it and announce him, it occurred to Dr. Warren that he would tell Mary that evening that if Mrs. Jameson had been the heroine of any unconventional domestic drama it was an unmistakable fact that Jane Cupp would have "felt it her duty as a young woman to leave this day month, if you please, ma'am," quite six months ago. And there she was, in a neat gown and apron,—evidently a fixture because she liked her place,—her decent young face full of sympathetic interest.

The day was dull and cold, but the front room was warm and made cheerful by fire. Mrs. Jameson was sitting at a writing-table. There were letters before her, and she seemed to have been re-reading them. She did not any longer bloom with normal health. Her face was a little dragged, and the first thing he noted in the eyes she lifted to him was that they were bewildered.

"She has had a shock," he thought. "Poor woman!"

He began to talk to her about herself with the kindly perception which was inseparable from him. He wondered if the time had not come when she would confide in him. Her shock, whatsoever it had been, had left her in the position of a woman wholly at a loss to comprehend what had occurred. He saw this in her ingenuous troubled face. He felt as if she was asking herself what she should do. It was not unlikely that presently she would ask him what she should do. He had been asked such things before by women, but they usually added trying detail accompanied by sobs, and appealed to his chivalry for impossible aid. Sometimes they implored him to go to people and use his influence.

Emily answered all his questions with her usual sweet, good sense. She was not well. Yesterday she had fainted.

"Was there any disturbing reason for the faint?" he inquired.

"It was because I was—very much disappointed," she answered, hesitating. "I had a letter which—It was not what I expected."

She was thinking desperately. She could understand nothing. It was not explainable that what she had written did not matter at all, that James should have made no reply.

"I was awake all night," she added.

"That must not go on," he said.

"I was thinking—and thinking," nervously.

"I can see that," was his answer.

Perhaps she ought to have courage to say nothing. It might be safer. But it was so lonely not to dare to ask anyone's advice, that she was getting frightened. India was thousands and thousands of miles away, and letters took so long to come and go. Anxiety might make her ill before she could receive a reply to a second letter. And perhaps now in her terror she had put herself into a ridiculous position. How could she send for Lady Maria to Mortimer Street and explain to her? She realised also that her ladyship's sense of humour might not be a thing to confide in safely.

Warren's strong, amiable personality was good for her. It served to aid her to normal reasoning. Though she was not aware of the fact, her fears, her simplicity, and her timorous adoration of her husband had not allowed her to reason normally in the past. She had been too anxious and too much afraid.

Her visitor watched her with great interest and no little curiosity. He himself saw that her mood was not normal. She did not look as poor Mrs. Jerrold had looked, but she was not in a normal state.

He made his visit a long one purposely. Tea was brought up, and he drank it with her. He wanted to give her time to make up her mind about him. When at last he rose to go away, she rose also. She looked nervously undecided, but let him go towards the door.

Her move forward was curiously sudden.

"No, no," she said. "Please come back. I—oh!—I really think I ought to tell you."

He turned towards her, wishing that Mary were with him. She stood trying to smile, and looking so entirely nice and well-behaved even in her agitation.

"If I were not so puzzled, or if there was *anybody*—" she said. "If you could only advise me; I must—I *must* keep safe."

"There is something you want to tell me?" he said quietly.

"Yes," she answered. "I am so anxious, and I am sure it must be bad for one to be anxious always. I have not dared to tell anyone. My name is not Mrs. Jameson, Dr. Warren. I am—I am Lady Walderhurst."

He barely managed to restrain a start. He was obliged to admit to himself that he had not thought of anything like this. But Mary had been right.

Emily blushed to her ears with embarrassment. He did not believe her.

"But I *am* really," she protested. "I *really* am. I was married last year. I was Emily Fox-Seton. Perhaps you remember."

She was not flighty or indignant. Her frank face was only a little more troubled than it had been before. She looked straight into his eyes without a doubt of his presently believing her. Good heavens! if—

She walked to the writing-table and picked up a number of letters. They were all stamped with the same seal. She brought them to him almost composedly.

"I ought to have remembered how strange it would sound," she said in her amenable voice. "I hope I am not doing wrong in speaking. I hope you won't mind my troubling you. It seemed as if I *couldn't* bear it alone any longer."

After which she told him her story.

* * * * *

The unadorned straightforwardness of the relation made it an amazing thing to hear, even more amazing than it would have been made by a more imaginative handling. Her obvious inability to cope with the unusual and villainous, combined with her entire willingness to obliterate herself in any manner in her whole-souled tenderness for the one present object of her existence, were things a man could not be unmoved by, even though experience led him to smile at the lack of knowledge of the world which had left her without practical defence. Her very humbleness and candour made her a drama in herself.

"Perhaps I was wrong to run away. Perhaps only a silly woman would have done such a queer, unconventional thing. But I could think of nothing else so likely to be quite safe, until Lord Walderhurst could advise me. And when his letter came yesterday, and he did not speak of what I had said—" Her voice quite failed her.

"Captain Osborn has detained your letter. Lord Walderhurst has not seen it."

Life began to come back to her. She had been so horribly bewildered as to think at moments that perhaps it might be that a man who was very much absorbed in affairs—

"The information you sent him is the most important, and moving, a man in his position could receive."

"Do you think so, *really*?" She lifted her head with new courage and her colour returned.

"It is impossible that it should be otherwise. It is, I assure you, *impossible*, Lady Walderhurst."

"I am so thankful," she said devoutly. "I am so *thankful* that I have told you."

Anything more touching and attractive than her full eyes and her grown-up child's smile he felt he had never seen.

CHAPTER TWENTY TWO

The attack of fever which had seemed to begin lightly for Lord Walderhurst assumed proportions such as his medical man had not anticipated. His annoyance at finding his duties interfered with fretted him greatly. He was not, under the circumstances, a good patient, and, partly as a result of his state of mind, he began, in the course of a few weeks, to give his doctors rather serious cause for anxiety. On the morning following Emily's confession to Dr. Warren she had received a letter from her husband's physician, notifying her of his new anxieties in connection with his patient. His lordship required extreme care and absolute freedom from all excitement. Everything which medical science and perfect nursing could do would be done. The writer asked Lady Walderhurst's collaboration with him in his efforts at keeping the invalid as far as possible in unperturbed spirits. For some time it seemed probable that letter writing and reading would be out of the question, but if, when correspondence might be resumed, Lady Walderhurst would keep in mind the importance of serenity to the convalescent, the case would have all in its favour. This, combined with expressions of sympathetic encouragement and assurances that the best might be hoped for, was the gist of the letter. When Dr. Warren arrived, Emily handed the epistle to him and watched him as he read it.

"You see," she said when he looked up, "that I did not speak too soon. Now I shall have to trust to you for everything. I could *never* have borne it *all* by myself. Could I?"

"Perhaps not," thinking it over; "but you are very brave."

"I don't think I'm brave," thinking it over on her own part, "but it seemed as if there were things I *must* do. But now you will advise me."

She was as biddable as a child, he told his wife afterwards, and that a woman of her height and carriage should be as biddable as she might have been at six years old, was an effective thing.

"She will do anything I tell her, she will go anywhere I advise. I advise that she shall go to her husband's house in Berkeley Square, and that together you and I will keep unobtrusive guard over her. All is quite simple, really. All would have been comparatively simple at the outset, if she had felt sure enough of her evidence to dare to confide in some practical person. But she was too uncertain and too much afraid of scandal, which might annoy her husband. She is deeply in awe of Lord Walderhurst and deeply in love with him."

"When one realises how unnecessary qualities and charms seem to be to the awakening of the tender emotion, it is rather dull, perhaps, to ask why. Yet one weakly asks it," was Mrs. Warren's summation.

"And one cannot supply the answer. But the mere devotion itself in this nice creature is a thing to be respected. She will control even her anxieties and reveal nothing while she writes her cheerful letters, as soon as she is allowed to write them."

"Lord Walderhurst will be told nothing?"

"Nothing until his recovery is complete. Now that she has made a clean breast of everything to me and given herself into my hands, I believe that she finds a sentimental pleasure in the thought of keeping her secret until he returns. I will confess to you, Mary, that I think that she has read of and tenderly sympathised with heroines who have done the like before. She does not pose to herself as a heroine, but she dwells affectionately on ingenuous mental pictures of what Lord Walderhurst will say. It is just as well that it should be so. It is better for her than fretting would be. Experience helped me to gather from the medical man's letter that his patient is in no condition to be told news of any kind, good or bad."

The house in Berkeley Square was reopened. Lady Walderhurst returned to it, as it was understood below stairs, from a visit to some German health resort. Mrs. Cupp and Jane returned with her. The wife of her physician in attendance was with her a great deal. It was most unfortunate for her ladyship that my lord was detained in India by illness.

The great household, having presented opened shutters to the world, went on in the even tenor of its way. There brooded over it, however, a sort of hushed dignity of atmosphere. The very housemaids wore an air of grave discretion. Their labours assumed the proportions of confidential interested service, in which they felt a private pride. Not one among them had escaped becoming attached to Lady Walderhurst.

Away from Palstrey, away from Mortimer Street, Emily began to find reality in the fact that everything had already become quite simple, after all. The fine rooms looked so well ordered and decent in a stately way. Melodramatic plotting ceased to exist as she looked at certain dignified sofas and impressive candelabra. Such things became even more impossible than they had become before the convincingness of the first floor front bedroom in Mortimer Street, She began to give a good deal of thought to the summer at Mallowe. There was an extraordinary luxury in living again each day of it, the morning when she had taken the third-class carriage which provided her with hot, labouring men in corduroys as companions, that fleeting moment when the tall man with the square face had passed the carriage and looked straight through her without seeming to see her at all. She sat and smiled tenderly at the mere reminiscent thought. And then the glimpse of him as he got into the high phaeton at the station; and the moment when Lady Maria had exclaimed "There's Walderhurst," and he had come swinging with his leisurely step across the lawn. And he had scarcely seemed to see her then, or notice her really when they met, until the morning he had joined her as she gathered the roses and had talked to her about Lady Agatha. But he had actually been noticing her a little even from the first—he had been thinking about her a little all the time. And how far she had been from guessing it when she had talked to Lady Agatha, how pleased she had been the morning of the rose gathering when he had seemed interested only in Agatha's self! She always liked to recall, however, the way in which he had asked the few questions about her own affairs. Her simplicity never wearied of the fascination of the way in which he had looked at her, standing on the pathway, with that delightful non-committal fixing of her with the monocle when she had said:

"People *are* kind. You see, I have nothing to give, and I always seem to be receiving."

And he had gazed at her quite unmovedly and answered only:

"What luck!"

But since then he had mentioned this moment as one of those in which he had felt that he might want to marry her, because she was so unconscious of the fact that she gave much more to everybody than she received, that she had so much to give and was totally unaware of the value of her gifts.

"His thoughts of me are so *beautiful* very often," was her favourite reflection, "though he always has that composed way of saying things. What he says seems more *valuable*, because he is like that."

In truth, his composed way of saying things it was which seemed to her incomparable. Even when, without understanding its own longing for a thing it lacked, her heart had felt itself a little unsustained she had never ceased to feel the fascination of his entire freedom from any shadow of interest in the mental attitude of others towards himself. When he stood and gazed at people through the glass neatly screwed into his eye, one felt that it was he whose opinion was of importance, not the other person's. Through sheer chill imperviousness he seemed entirely detached from the powers of criticism. What people said or thought of his fixed opinion on a subject was not of the least consequence, in fact did not exist; the entities of the persons who cavilled at such opinions themselves ceased to exist, so far as he was concerned. His was the immovable temperament. He did not snub people: he cut the cord of mental communication with them and dropped them into space. Emily thought this firmness and reserved dignity, and quailed before the thought of erring in such a manner as would cause him to so send her soul adrift. Her greatest terror during the past months had been the fear of making him ridiculous, of putting him in some position which might annoy him by objectionable publicity.

But now she had no further fears, and could wait in safety and dwell in peace upon her memories and her hopes. She even began to gain a kind of courage in her thoughts of him.

The atmosphere of the Berkeley Square mansion was good for her. She had never felt so much its mistress before the staff of servants of whose existence she was the centre, who so plainly served her with careful pleasure, who considered her least wish or inclination as a royal command, increased her realisation of her security and power. The Warrens, who understood the dignity and meaning of mere worldly facts her nature did not grasp, added subtly to her support. Gradually she learned to reveal herself in simple talk to Mrs. Warren, who found her, when so revealed, a case more extraordinary than she had been when enshrouded in dubious mystery.

"She is absolutely delicious," Mrs. Warren said to her husband. "That an adoration such as hers could exist in the nineteenth century is—"

"Almost degenerate," he laughed.

"Perhaps it is regenerate," reflecting. "Who knows! Nothing earthly, or heavenly, would induce me to cast a doubt upon it. Seated opposite to a portrait of her James, I hear her opinions of him, when she is not in the least aware of what her simplest observation conveys. She does not know that she is including him when she is talking of other things, that one sees that while she is too shy to openly use his name much, the very breath of her life is a reference to him. Her greatest bliss at present is to go unobtrusively into his special rooms and sit there dwelling upon his goodness to her."

In fact Emily spent many a quiet hour in the apartments she had visited on the day of her farewell to her husband. She was very happy there. Her soul was uplifted by her gratitude for the peace she had reached. The reports of Lord Walderhurst's physician were never alarming and generally of a reassuring nature. But she knew that he must exercise great caution, and that time must elapse before he could confront his return voyage. He would come back as soon as was quite safe. And in the meantime her world held all that she could desire, lacking himself.

Her emotion expressed itself in her earnest performance of her reverent daily devotions. She read many chapters of the Bible, and often sat happily absorbed in the study of her Book of Common Prayer. She found solace and happiness in such things, and spent her Sunday mornings, after the ringing of the church bells, quite alone in Walderhurst's study, following the Service and reading the Collects and Lessons. The room used to seem so beautifully still, even Berkeley Square wearing its church-hour aspect suggested devout aloofness from worldly things.

"I sit at the window and *think*," she explained to Mrs. Warren. "It is so nice there."

She wrote her letters to India in this room. She did not know how far the new courage in her thoughts of her husband expressed itself in these letters. When Walderhurst read them, however, he felt a sense of change in her. Women were sometimes spoken of as "coming out amazingly." He began to feel that Emily was, in a measure at least, "coming out." Perhaps her gradually increasing feeling of accustomedness to the change in her life was doing it for her. She said more in her letters, and said it in a more interesting way. It was perhaps rather suggestive of the development of a girl who was on the verge of becoming a delightful sort of woman.

Lying upon his back in bed, rendered, it may be, a trifle susceptible by the weakness of slow convalescence, he found a certain habit growing upon him—a habit of reading her letters several times, and of thinking of her as it had not been his nature to think of women; also he slowly awakened to an interest in the arrival of the English mails. The letters actually raised his spirits and had an excellent physical effect. His doctor always found him in good condition after he had heard from his wife.

"Your letters, my dear Emily," Walderhurst once wrote, "are a great pleasure to me. You are to-day exactly as you were at Mallowe,—the creature of amiable good cheer. Your comfort stimulates me."

"How *dear*, how *dear*?" Emily cried to the silence of the study, and kissed the letter with impassioned happiness.

Lady Maria Bayne

The next epistle went even farther. It absolutely contained "things" and referred to the past which it was her joy to pour libations before in secret thought. When her eye caught the phrase "the days at Mallowe" in the middle of a sheet, she was almost frightened at the rush of pleasure which swept over her. Men who were less aloof from sentimental moods used such phrases in letters, she had read and heard. It was almost as if he had said "the dear old days at Mallowe" or "the happy days at Mallowe," and the rapture of it was as much as she could bear.

"I cannot help remembering as I lie here," she read in actual letters as she went on, "of the many thoughts which passed through my mind as I drove over the heath to pick you up. I had been watching you for days. I always liked particularly your clear, large eyes. I recall trying to describe them to myself and finding it difficult. They seemed to me then to resemble something between the eyes of a very nice boy and the eyes of a delightful sheep-dog. This may not appear so romantic a comparison as it really is."

Emily began most softly and sweetly to cry. Nothing more romantic could she possibly have imagined.

"I thought of them in spite of myself as I drove across the moor, and I could scarcely express to you how angry I was at Maria. It seemed to me that she had brutally imposed on you only because she had known she might impose on a woman with such a pair of eyes. I was angry and sentimental at one and the same time. And to find you sitting by the wayside, absolutely worn out with fatigue and in tears, moved me really more than I had anticipated being moved. And when you mistook my meaning and stood up, your nice eyes looking into mine in such ingenuous appeal and fear and trouble, I have never forgotten it, my dear, and I never shall."

His mood of sentiment did not sit easily upon him, but it meant a real and interesting quite human thing.

Emily sat alone in the room and brooded over it as a mother might brood over a new-born child. She was full of tremulous bliss, and, dwelling with reverent awe upon the wonder of great things drawing nearer to her every hour, wept for happiness as she sat.

* * * * *

The same afternoon Lady Maria Bayne arrived. She had been abroad taking, in no dull fashion, various "cures," which involved drinking mineral waters while promenading to the sounds of strains of outdoor music, and comparing symptoms wittily with friends equal to amazing repartee in connection with all subjects.

Dr. Warren was an old acquaintance, and as he was on the point of leaving the house as she entered it she stopped to shake hands with him.

"It's rather unfortunate for a man when one can only be glad to see him in the house of an enemy."

She greeted him with, "I must know what you are doing here. It's not possible that Lady Walderhurst is fretting herself into fiddle-strings because her husband chooses to have a fever in India."

"No, she is behaving beautifully in all respects. May I have a few minutes' talk with you, Lady Maria, before you see her?"

"A few minutes' talk with me means something either amusing or portentous. Let us walk into the morning-room."

She led the way with a rustle or silk petticoats and a suggestion of lifted eyebrows. She was inclined to think that the thing sounded more portentous than amusing. Thank Heaven! it was not possible for Emily to have involved herself in annoying muddles. She was not that kind of woman.

When she came out of the room some twenty minutes later she did not look quite like herself. Her smart bonnet set less well upon her delicate little old face, and she was agitated and cross and pleased.

"It was ridiculous of Walderhurst to leave her," she was saying. "It was ridiculous of her not to order him home at once. It was exactly like her,—dear and ridiculous."

In spite of her agitation she felt a little grotesque as she went upstairs to see Emily,—grotesque, because she was obliged to admit to herself that she had never felt so curiously excited in her life. She felt as she supposed women did when they allowed themselves to shed tears through excitement; not that she was shedding tears, but she was "upset," that was what she called it.

As the door opened Emily rose from a chair near the fire and came slowly towards her, with an awkward but lovely smile.

Lady Maria made a quick movement forward and caught hold of both her hands.

"My good Emily," she broke forth and kissed her. "My excellent Emily," and kissed her again. "I am completely turned upside down. I never heard such an insane story in my life. I have seen Dr. Warren. The creatures were mad."

"It is all over," said Emily. "I scarcely believe it was true now."

Lady Maria being led to a sofa settled herself upon it, still wearing her complex expression of crossness, agitation, and pleasure.

"I am going to stay here," she said, obstinately. "There shall be no more folly. But I will tell you that they have gone back to India. The child was a girl."

"It was a girl?"

"Yes, absurdly enough."

"Oh," sighed Emily, sorrowfully. "I'm *sure* Hester was *afraid* to write to me."

"Rubbish!" said Lady Maria. "At any rate, as I remarked before, I am going to stay here until Walderhurst comes back. The man will be quite mad with gratified vanity."

CHAPTER TWENTY THREE

It was a damp and depressing day on which Lord Walderhurst arrived in London. As his carriage turned into Berkeley Square he sat in the corner of it rather huddled in his travelling-wraps and looking pale and thin. He was wishing that London had chosen to show a more exhilarating countenance to him, but he himself was conscious of being possessed by something more nearly approaching a mood of eagerness than he remembered experiencing at any period of his previous existence. He had found the voyage home long, and had been restless. He wanted to see his wife. How agreeable it would be to meet, when he looked across the dinner-table, the smile in her happy eyes. She would grow pink with pleasure, like a girl, when he confessed that he had missed her. He was curious to see in her the changes he had felt in her letters. Having time and opportunities for development, she might become an absolutely delightful companion. She had looked very handsome on the day of her presentation at Court. Her height and carriage had made her even impressive. She was a woman, after all, to be counted on in one's plans.

But he was most conscious that his affection for her had warmed. A slight embarrassment was commingled with the knowledge, but that was the natural result of his dislike to the sentimental. He had never felt a shadow of sentiment for Audrey, who had been an extremely light, dry, empty-headed person, and he had always felt she had been adroitly thrust upon him by their united families. He had not liked her, and she had not liked him. It had been very stupidly trying. And the child had not lived an hour. He had liked Emily from the first, and now—It was an absolute truth that he felt a slight movement in the cardiac region when the carriage turned into Berkeley Square. The house would look very pleasant when he entered it. Emily would in some subtle way have arranged that it should wear a festal, greeting air. She had a number of nice, little feminine emotions about bright fires and many flowers. He could picture her childlike grown-up face as it would look when he stepped into the room where they met.

Some one was ill in Berkeley Square, evidently very ill. Straw was laid thick all along one side of it, depressing damp, fresh straw, over which the carriage rolled with a dull drag of the wheels.

It lay before the door of his own house, he observed, as he stepped out. It was very thickly scattered. The door swung open as the carriage stopped. Crossing the threshold, he glanced at the face of the footman nearest to him. The man looked like a mute at a funeral, and the expression was so little in accord with his mood that he stopped with a feeling of irritation. He had not time to speak, however, before a new sensation arrested his attention,—a faint odour which filled the place.

"The house smells like a hospital," he exclaimed, in great annoyance. "What does it mean?"

The man he addressed did not answer. He turned a perturbed awkward face to his superior in rank, an older man, who was house steward.

In the house of mortal pain or death there is but one thing more full of suggestion than the faint smell of antiseptics,—the gruesome, cleanly, unpleasant odour,—that is, the unnatural sound of the whispering of hushed voices. Lord Walderhurst turned cold, and felt it necessary to stiffen his spine when he heard his servant's answer and the tone in which it was made.

"Her ladyship, my lord—her ladyship is very low. The doctors do not leave her."

"Her ladyship?"

The man stepped back deferentially. The door of the morning-room had been opened, and old Lady Maria Bayne stood on the threshold. Her worldly air of elderly gaiety had disappeared. She looked a hundred. She was almost dilapidated. She had allowed to relax themselves the springs which held her together and ordinarily supplied her with sprightly movement.

"Come here!" she said.

When he entered the room, aghast, she shut the door.

"I suppose I ought to break it to you gently," she said shakily, "but I shall do no such thing. It's too much to expect of any woman who has gone through what I have during these last three days. The creature is dying; she may be dead now."

She sank on the sofa and began to wipe away pouring tears. Her old cheeks were pale and her handkerchief showed touches of rose-pink on its dampness. She was aware of their presence, but was utterly indifferent. Walderhurst stared at her haggard disorder and cleared his throat, finding himself unable to speak without doing so.

"Will you have the goodness to tell me," he said with weird stiffness, "what you are talking about?"

"About Emily Walderhurst," she answered. "The boy was born yesterday, and she has been sinking ever since. She cannot possibly last much longer."

"She!" he gasped, turning lead colour. "Cannot possibly last,—Emily?"

The wrench and shock were so unnatural that they reached that part of his being where human feeling was buried under selfishness and inhuman conventionality. He spoke, and actually thought, of Emily first.

Lady Maria continued to weep shamelessly.

"I am over seventy," she said, "and the last three days have punished me quite enough for anything I may have done since I was born. I have been in hell, too, James. And, when she could think at all, she has only thought of you and your miserable child. I can't imagine what is the matter with a woman when she can care for a man to such an extent. Now she has what she wants,—she's dying for you."

"Why wasn't I told?" he asked, still with the weird and slow stiffness.

"Because she was a sentimental fool, and was afraid of disturbing you. She ought to have ordered you home and kept you dancing attendance, and treated you to hysterics."

No one would have resented such a course of action more derisively than Lady Maria herself, but the last three days had reduced her to something like hysteria, and she had entirely lost her head.

"She has been writing cheerfully to me—"

"She would have written cheerfully to you if she had been seated in a cauldron of boiling oil, it is my impression," broke in her ladyship. "She has been monstrously treated, people trying to murder her, and she afraid to accuse them for fear that you would disapprove. You know you have a nasty manner, James, when you think your dignity is interfered with."

Lord Walderhurst stood clenching and unclenching his hands as they hung by his sides. He did not like to believe that his fever had touched his brain, but he doubted his senses hideously.

"My good Maria," he said, "I do not understand a word you say, but I must go and see her."

"And kill her, if she has a breath left! You will not stir from here. Thank Heaven! here is Dr. Warren."

The door had opened and Dr. Warren came in. He had just laid down upon the coverlet of a bed upstairs what seemed to be the hand of a dying woman, and no man like himself can do such a thing and enter a room without a singular look on his face.

People in a house of death inevitably whisper, whatsoever their remoteness from the sick-room. Lady Maria cried out in a whisper:

"Is she still alive?"

"Yes," was the response.

Walderhurst went to him.

"May I see her?"

"No, Lord Walderhurst. Not yet."

"Does that mean that it is not yet the last moment?"

"If that moment had obviously arrived, you would be called."

"What must I do?"

"There is absolutely nothing to be done but to wait. Brent, Forsythe, and Blount are with her."

"I am in the position of knowing nothing. I must be told. Have you time to tell me?"

They went to Walderhurst's study, the room which had been Emily's holy of holies.

"Lady Walderhurst was very fond of sitting here alone," Dr. Warren remarked.

Walderhurst saw that she must have written letters at his desk. Her own pen and writing-tablet lay on it. She had probably had a fancy for writing her letters to himself in his own chair. It would be like her to have done it. It gave him a shock to see on a small table a thimble and a pair of scissors.

"I ought to have been told," he said to Dr. Warren.

Dr. Warren sat down and explained why he had not been told.

As he spoke, interest was awakened in his mind by the fact that Lord Walderhurst drew towards him the feminine writing-tablet and opened and shut it mechanically.

"What I want to know," he said, "is, if I shall be able to speak to her. I should like to speak to her."

"That is what one most wants," was Dr. Warren's non-committal answer, "at such a time."

"You think I may not be able to make her understand?"

"I am very sorry. It is impossible to know."

"This," slowly, "is very hard on me."

"There is something I feel I must tell you, Lord Walderhurst." Dr. Warren kept a keen eye on him, having, in fact, felt far from attracted by the man in the past, and wondering how much he would be moved by certain truths, or if he would be moved at all. "Before Lady Walderhurst's illness, she was very explicit with me in her expression of her one desire. She begged me to give her my word, which I could not have done without your permission, that whatsoever the circumstances, if life must be sacrificed, it should be hers."

A dusky red shot through Walderhurst's leaden pallor.

"She asked you that?" he said.

"Yes. And at the worst she did not forget. When she became delirious, and we heard that she was praying, I gathered that she seemed to be praying to me, as to a deity whom she implored to remember her fervent pleading. When her brain was clear she was wonderful. She saved your son by supernatural endurance."

"You mean to say that if she had cared more for herself and less for the safety of the child she need not have been as she is now?"

Warren bent his head.

Lord Walderhurst's eyeglass had been dangling weakly from its cord. He picked it up and stuck it in his eye to stare the doctor in the face. The action was a singular, spasmodic, hard one. But his hands were shaking.

"By God!" he cried out, "if I had been here it should not have been so!"

He got up and supported himself against the table with the shaking hands.

"It is very plain," he said, "that she has been willing to be torn to pieces upon the rack to give me the thing I wanted. And now, good God in heaven, I feel that I would have strangled the boy with my own hands rather than lose her."

In this manner, it seemed, did a rigid, self-encased, and conventional elderly nobleman reach emotion. He looked uncanny. His stiff dignity hung about him in rags and tatters. Cold sweat stood on his forehead and his chin twitched.

"Just now," he poured forth, "I don't care whether there is a child or not. I want her—I care for nothing else. I want to look at her, I want to speak to her, whether she is alive or dead. But if there is a spark of life in her, I believe she will hear me."

Dr. Warren sat and watched him, wondering. He knew curious things of the human creature, things which most of his confrères did not know. He knew that Life was a mysterious thing, and that even a dying flame of it might sometimes be fanned to flickering anew by powers more subtle than science usually regards as applicable influences. He knew the nature of the half-dead woman lying on her bed upstairs, and he comprehended what the soul of her life had been,—her divinely innocent passion for a self-centred man. He had seen it in the tortured courage of her eyes in hours of mortal agony.

"Don't forget," she had said. "Our Father which art in Heaven. Don't let anyone forget. Hallowed be thy name."

The man, leaning upon his shaking hands before him, stood there, for these moments at least, a harrowed thing. Not a single individual of his acquaintance would have known him.

"I want to see her before the breath leaves her," he gave forth in a harsh, broken whisper. "I want to speak. Let me see her."

Dr. Warren left his chair slowly. Out of a thousand chances against her, might this one chance be for her,—the chance of her hearing, and being called back to the shores she was drifting from, by this stiff, conventional fellow's voice. There was no knowing the wondrousness of a loving human thing, even when its shackles were loosening themselves to set it free.

"I will speak to those in charge with me," he said. "Will you control every outward expression of feeling?"

"Yes."

Adjoining Lady Walderhurst's sleeping apartment was a small boudoir where the medical men consulted together. Two of them were standing near the window conversing in whispers.

Walderhurst merely nodded and went to wait apart by the fire. Ceremony had ceased to exist. Dr. Warren joined the pair at the window. Lord Walderhurst only heard one or two sentences.

"I am afraid that nothing, now, can matter—at any moment."

* * * * *

Those who do not know from experience what he saw when he entered the next room have reason to give thanks to such powers as they put trust in.

There ruled in the large, dim chamber an awful order and silence. The faint flickering of the fire was a marked sound. There was no other but a fainter and even more irregular one heard as one neared the bed. Sometimes it seemed to stop, then, with a weak gasp, begin again. A nurse in uniform stood in waiting; an elderly man sat on a chair at the bedside, listening and looking at his watch, something white and lifeless lying in his grasp,—Emily Walderhurst's waxen, unmoving hand. The odour of antiseptics filled the nostrils. Lord Walderhurst drew near. The speaking sign of the moment was that neither nurse nor doctor stirred.

Emily lay low upon a pillow. Her face was as bloodless as wax and was a little turned aside. The Shadow was hovering over it and touched her closed lids and the droop of her cheek and corners of her mouth. She was far, far away.

This was what Walderhurst felt first,—the strange remoteness, the lonely stillness of her. She had gone alone far from the place he stood in, and which they two familiarly knew. She was going, alone, farther still. As he stood and watched her closed eyes,—the nice, easily pleased eyes,—it was they themselves, closed on him and all prosaic things and pleasures, which filled him most strangely with that sense of her loneliness, weirdly enough, *hers*, not his. He was not thinking of himself but of her. He wanted to withdraw her from her loneliness, to bring her back.

He knelt down carefully, making no sound, stealthily, not removing his eyes from her strange, aloof face. He slowly dared to close his hand on hers which lay outside the coverlet. And it was a little chill and damp,—a little chill.

A power, a force which hides itself in human things and which most of them know not of, was gathering within him. He was warm and alive, a living man; his hand as it closed on the chill of hers was warm; his newly awakened being sent heat to it.

He whispered her name close to her ear.

"Emily!" slowly, "Emily!"

She was very far away and lay unmoving. Her breast scarcely stirred with the faintness of her breath.

"Emily! Emily!"

The doctor slightly raised his eyes to glance at him. He was used to death-bed scenes, but this was curious, because he knew the usual outward aspect of Lord Walderhurst, and its alteration at this moment suggested abnormal things. He had not the flexibility of mind which revealed to Dr. Warren that there were perhaps abnormal moments for the most normal and inelastic personages.

"Emily!" said his lordship, "Emily!"

He did not cease from saying it, in a low yet reaching whisper, at regular intervals, for at least half an hour. He did not move from his knees, and so intense was his absorption that the presence of those who came near was as nothing.

What he hoped or intended to do he did not explain to himself. He was of the order of man who coldly waves aside all wanderings on the subjects of occult claims. He believed in proven facts, in professional aid, in the abolition of absurdities. But his whole narrow being concentrated itself on one thing,—he wanted this woman back. He wanted to speak to her.

What power he unknowingly drew from the depths of him, what exquisite answering thing he reached at, could not be said. Perhaps it was only some remote and subtle turn of the tide of life and death which chanced to come to his aid.

"Emily!" he said again, after many times.

Dr. Warren at this moment met the lifted eyes of the doctor who was counting her pulse, and in response to his look went to him.

"It seems slightly stronger," Dr. Forsythe whispered.

The slow, faint breathing changed a shade; there was heard a breath slightly, very slightly deeper, less flickering, then another.

Lady Walderhurst slightly stirred.

"Remain where you are," whispered Dr. Warren to her husband, "and continue to speak to her. Do not alter your tone. Go on."

* * * * *

Emily Walderhurst, drifting out on a still, borderless, white sea, sinking gently as she floated, sinking in peaceful painlessness deeper and deeper in her drifting until the soft, cool water lapped her lips and, as she knew without fear, would soon cover them and her quiet face, hiding them for ever,—heard from far, very far away, across the whiteness floating about her, a faint sound which at first only fell upon the stillness without meaning. Everything but the silence had been left behind aeons ago. Nothing remained but the soundless white sea and the slow drifting and sinking as one swayed. It was more than sleep, this still peace, because there was no thought of waking to any shore.

But the far-off sound repeated itself again, again, again and again, monotonously. Something was calling to Something. She was so given up to the soft drifting that she had no thoughts to give, and gave none. In drifting so, one did not think—thought was left in the far-off place the white sea carried one from. She sank quietly a little deeper and the water touched her lip. But Something was calling to Something, something was calling something to come back. The call was low, low and strange, so regular and so unbroken and insistent, that it arrested, she knew not what. Did it arrest the floating and the swaying in the enfolding sea? Was the drifting slower? She could not rouse herself to think, she wanted to go on. Did she no longer feel the water lapping against her lip? Something was calling to Something still. Once, aeons ago, before the white sea had borne her away, she would have understood.

"Emily, Emily, Emily!"

Yes, once she would have known what the sound meant. Once it had meant something, a long time ago. It had even now disturbed the water, and made it cease to lap so near her lip.

* * * * *

It was at this moment that one doctor had raised his eyes to the other, and Lady Walderhurst had stirred.

When Walderhurst left his place beside his wife's bed, Dr. Warren went with him to his room. He made him drink brandy and called his man to him. "You must remember," he said, "that you are an invalid yourself."

"I believe," was the sole answer, given with an abstracted knitting of the brows,—"I believe that in some mysterious way I have made her hear me."

Dr. Warren looked grave. He was a deeply interested man. He felt that he had been looking on at an almost incomprehensible thing.

"Yes," was his reply. "I believe that you have."

About an hour later Lord Walderhurst made his way downstairs to the room in which Lady Maria Bayne sat. She still looked a hundred years old, but her maid had redressed her toupee, and given her a handkerchief neither damp nor tinted with rubbed-off rouge. She looked at her relative a shade more leniently, but still addressed him with something of the manner of a person undeservedly chained to a malefactor. Her irritation was not modified by the circumstance that it was extremely difficult to be definite in the expression of her condemnation of things which had made her hideously uncomfortable. Having quite approved of his going to India in the first place, it was not easy to go thoroughly into the subject of the numerous reasons why a man of his years and responsibilities ought to have realised that it was his duty to remain at home and take care of his wife.

"Incredible as it seems," she snapped, "the doctors *think* there is a slight change, for the better."

"Yes," Walderhurst answered.

He leaned against the mantel and gazed into the fire.

"She will come back," he added in a monotone.

Lady Maria stared at him. She felt that the man was eerie, Walderhurst, of all men on earth!

"Where do you think she has been?" She professed to make the inquiry with an air of reproof.

"How should one know?" rather with the old stiffness. "It is impossible to tell."

Lady Maria Bayne was not the person possessing the temperament to incline him to explain that, wheresoever the outer sphere might be to which the dying woman had been drifting, he had been following her, as far as living man could go.

The elderly house steward opened the door and spoke in the hollow whisper.

"The head nurse wished to know if your ladyship would be so good as to see Lord Oswyth before he goes to sleep."

Walderhurst turned his head towards the man. Lord Oswyth was the name of his son. He felt a shock.

"I will come to the nursery," answered Lady Maria. "You have not seen him yet?" turning to Walderhurst.

"How could I?"

"Then you had better come now. If she becomes conscious and has life enough to expect anything, she will expect you to burst forth into praises of him. You had better at least commit to memory the colour of his eyes and hair. I believe he has two hairs. He is a huge, fat, overgrown thing with enormous cheeks. When I saw his bloated self-indulgent look yesterday, I confess I wanted to slap him."

Her description was not wholly accurate, but he was a large and robust child, as Walderhurst saw when he beheld him.

From kneeling at the pillow on which the bloodless statue lay, and calling into space to the soul which would not hear, it was a far cry to the warmed and lighted orris-perfumed room in which Life had begun.

There was the bright fire before which the high brass nursery fender shone. There was soft linen hanging to be warmed, there was a lace-hung cradle swinging in its place, and in a lace-draped basket silver and gold boxes and velvet brushes and sponges such as he knew nothing about. He had not been in such a place before, and felt awkward, and yet in secret abnormally moved, or it seemed abnormally to him.

Two women were in attendance. One of them held in her arms what he had come to see. It was moving slightly in its coverings of white. Its bearer stood waiting in respectful awe as Lady Maria uncovered its face.

"Look at it," she said, concealing her relieved elation under a slightly caustic manner. "How you will relish the situation when Emily tells you that he is like you, I can't be as sure as I should be of myself under the same circumstances."

Walderhurst applied his monocle and gazed for some moments at the object before him. He had not known that men experienced these curiously unexplainable emotions at such times. He kept a strong hold on himself.

"Would you like to hold him?" inquired Lady Maria. She was conscious of a benevolent effort to restrain the irony in her voice.

Lord Walderhurst made a slight movement backward.

"I—I should not know how," he said, and then felt angry at himself. He desired to take the thing in his arms. He desired to feel its warmth. He absolutely realised that if he had been alone with it, he should have laid aside his eyeglass and touched its cheek with his lips.

Two days afterwards he was sitting by his wife's pillow, watching her shut lids, when he saw them quiver and slowly move until they were wide open. Her eyes looked very large in her colourless, more sharply chiselled face. They saw him and him only, as light came gradually into them. They did not move, but rested on him. He bent forward, almost afraid to stir.

He spoke to her as he had spoken before.

"Emily!" very low, "Emily!"

Her voice was only a fluttering breath, but she answered.

"It—was—you!" she said.

CHAPTER TWENTY FOUR

Such individuals as had not already thought it expedient to gradually loosen and drop the links of their acquaintance with Captain Alec Osborn did not find, on his return to his duties in India, that the leave of absence spent in England among his relatives had improved him. He was plainly consuming enormous quantities of brandy, and was steadily going, physically and mentally, to seed. He had put on flesh, and even his always dubious good looks were rapidly deserting him. The heavy young jowl looked less young and more pronounced, and he bore about an evil countenance.

"Disappointment may have played the devil with him," it was said by an elderly observer; "but he has played the devil with himself. He was a wrong'un to begin with."

When Hester's people flocked to see her and hear her stories of exalted life in England, they greeted her with exclamations of dismay. If Osborn had lost his looks, she also had lost hers. She was yellow and haggard, and her eyes looked over-grown. She had not improved in the matter of temper, and answered all effusive questions with a dry, bitter little smile. The baby she had brought back was a puny, ugly, and tiny girl. Hester's dry, little smile when she exhibited her to her relations was not pretty.

"She saved herself disappointment by being a girl," she remarked. "At all events, she knows from the outset that no one can rob her of the chance of being the Marquis of Walderhurst."

It was rumoured that ugly things went on in the Osborn bungalow. It was known that scenes occurred between the husband and wife which were not of the order admitted as among the methods of polite society. One evening Mrs. Osborn walked slowly down the Mall dressed in her best gown and hat, and bearing on her cheek a broad, purpling mark. When asked questions, she merely smiled and made no answer, which was extremely awkward for the well-meaning inquirer.

The questioner was the wife of the colonel of the regiment, and when the lady related the incident to her husband in the evening, he drew in his breath sharply and summed the situation up in a few words.

"That little woman," he said, "lives every day through twenty-four hours of hell. One can see it in her eyes, even when she professes to smile at the brute for decency's sake. The awfulness of a woman's forced smile at the devil she is tied to, loathing him and bearing in her soul the thing, blood itself could not wipe out. Ugh! I've seen it once before, and I recognised it in her again. There will be a bad end to this."

There probably would have been, with the aid of unlimited brandy and unrestrained devil, some outbreak so gross that the social laws which rule men who are "officers and gentlemen" could not have ignored or overlooked it. But the end came in an unexpected way, and Osborn was saved from open ignominy by an accident.

On a certain day when he had drunk heavily and had shut Hester up with him for an hour's torture, after leaving her writhing and suffocating with sobs, he went to examine some newly bought firearms. In twenty minutes it was he who lay upon the floor writhing and suffocating, and but a few minutes later he was a dead man. A charge from a gun he had believed unloaded had finished him.

* * * * *

Lady Walderhurst was the kindest of women, as the world knew. She sent for little Mrs. Osborn and her child, and was tender goodness itself to them.

Hester had been in England four years, and Lord Oswyth had a brother as robust as himself, when one heavenly summer afternoon, as the two women sat on the lawn drinking little cups of tea, Hester made a singular revelation, and made it without moving a muscle of her small countenance.

"I always intended to tell you, Emily," she began quietly, "and I will tell you now."

"What, dear?" said Emily, holding out to her a plate of tiny buttered scones. "Have some of these nice, little hot ones."

"Thank you." Hester took one of the nice, little hot ones, but did not begin to eat it. Instead, she held it untouched and let her eyes rest on the brilliant flower terraces spread out below. "What I meant to tell you was this. The gun was not loaded, the gun Alec shot himself with, when he laid it aside."

Emily put down her tea-cup hastily.

"I saw him take out the charge myself two hours before. When he came in, mad with drink, and made me go into the room with him, Ameerah saw him. She always listened outside. Before we left The Kennel Farm, the day he tortured and taunted me until I lost my head and shrieked out to him that I had told you what I knew, and had helped you to go away, he struck me again and again. Ameerah heard that. He did it several times afterwards, and she always knew. She always intended to end it in some way. She knew how drunk he was that last day, and—It was she who went in and loaded the gun while he was having his scene with me. She knew he would go and begin to pull the things about without having the sense to know what he was doing. She had seen him do it before. I know it was she who put the load in. We have never uttered a word to each other about it, but I know she did it, and that she knows I know. Before I married Alec, I did not understand how one human being could kill another. He taught me to understand, quite. But I had not the courage to do it myself. Ameerah had."

And while Lady Walderhurst sat gazing at her with a paling face, she began quietly to eat the little buttered scone.

THE END

BIBLIOBAZAAR

The essential book market!

Did you know that you can get any of our titles in large print?

Did you know that we have an ever-growing collection of books in many languages?

**Order online:
www.bibliobazaar.com**

Find all of your favorite classic books!

Stay up to date with the latest government reports!

At BiblioBazaar, we aim to make knowledge more accessible by making thousands of titles available to you- *quickly and affordably*.

Contact us:
BiblioBazaar
PO Box 21206
Charleston, SC 29413